# Tracks in the Forest

KEN DRUSHKA—HANNU KONTTINEN

# TRACKS *in the Forest*

## The Evolution of Logging Machinery

TIMBERJACK GROUP • HELSINKI

Drushka, Ken — Konttinen, Hannu

**Tracks in the Forest**

The Evolution of Logging Machinery

Includes bibliographical and photo source reference and index.

Published by Timberjack Group Oy
Fabianinkatu 9 A
00100 Helsinki
Finland

Graphic design: Jorma Hinkka, Graafiset Neliöt Oy
Reproduction: Repromestarit Oy
Printed in Finland by Otava Printing Works, Keuruu 1997



ISBN 952-90-8616-4

As Timberjack celebrates its 50th anniversary,
this book is our tribute to today's
forest professionals and to the inventors and innovators
who have contributed to the development of
timber harvesting,
making it safer, more reliable and more productive.

# Contents

**10** Foreword

**15** Introduction

**21** From Muscles to Machines

**45** The Birth of Mechanization

**69** Post-war Recovery

**93** The Rubber Tire Revolution Begins

**117** Searching for the Right Combinations

**143** Improving on the Details

**165** The Return of the Processors

**193** The Modern Era

**219** Epilogue

**225** Appendix

- *227 Glossary*
- *229 Wood Measurement Equivalents*
- *230 Oral History and Story Sources*
- *232 Photo Sources*
- *233 Index*
- *241 Global Forest Resources – 1994*
- *247 Global Forest Machine Markets 1956–1996*
- *248 Highlights of Forest Machine Evolution*
- *251 A Short History of Timberjack's Evolution*
- *252 Sources and Acknowledgments*
- *253 The Authors*
- *254 Authors' Comments*

309
226
228
227
229

Forest machines have mainly been developed in countries of Northern coniferous forests. In these areas, forestry and related industries form the bases for national economies. Lots of attention is paid on the development of silviculture methods, as well. In the picture trees are marked for thinning in the Punkaharju test area of Finnish Forest Research Institute in 1925.

# A real-life adventure

For 50 years, Timberjack machines and equipment have been harvesting timber. We are marking this milestone with 'Tracks in the Forest', a tribute to the people who have contributed to the mechanization of forest work and the evolution of timber harvesting. This is the story of that adventure.

Early attempts at mechanized timber harvesting date back nearly a century, but the development of forest machines has largely occurred during the past 50 years. Timberjack has not been alone in this process. The main credit goes to the forest industry and to the forest professionals who came before us. Many companies — several of which are still our main competitors — also deserve credit.

Once we came up with the idea, the publication of this book quickly became a matter of conscience. We were determined to leave a chronicle of forest machine development for future generations. Several decades ago, the goal of mechanization was to get timber out of the forest as cheaply as possible. Machines were designed accordingly, and timber was dragged from the forest without regard for the traces left behind. At this time, the environment was an unknown concept, one that was never discussed. Decisions about forest machines, harvesting, forest management and regeneration were made with little or no concern for the environment.

Recent decades have seen drastic changes in the expectations and needs that forest machines must meet. This has definitely been a wise course of development. Today's forest machines can leave as little trace as a woodcutter or horse — as long as the work is well planned and implemented by professionals. By the same token, forest machines can leave deep scars if the operators are careless or inexperienced. Nonetheless, forest machine designers have given us clear cause for optimism. Vast improvements have been made in the environmental features of these machines and the operators' working conditions. The good news for the general public is that the forest machine sector recognizes environmental issues and is focusing as never before on forest-friendly product development and technology. Progress in the area of low-impact harvesting depends less on the forest machines themselves and more on forest use policy, forest management guidelines, and the planning and implementation of logging sites. The forest machine sector is capable of producing the kind of equipment needed to achieve the desired results.

Forestry work is now largely mechanized throughout the world. In those areas which have yet to mechanize, machines will soon replace axes and draft animals. There is no turning back to the heavy, exhausting and dangerous job of manual forest work. Instead of decreasing, global harvesting volumes are set to increase, and the same is true for the mechanization and automation of forest work. At the same time, the quality of the work is improving. For this reason, it pays to focus on the development of even better methods, and to produce timber following commonly accepted guidelines in areas suitable for harvesting. This will ensure that we have enough forests remaining for conservation and recreational use.

Measured by the earth's clock, fifty years is barely an instant in time. This short era does not warrant an official history. Instead, our intention was to write about the real-life story of the industry. We also wanted to introduce the inventors and machine shops with deep roots in the forest. The birth, growth and maturation of forest machine companies, complete with their successes and failures, were also a topic of interest. With this in mind, we were not looking for industry insiders — either engineers or forest professionals — to write the book. It takes some distance, after all, to see the forest for the trees. We wanted writers who represented both the Scandinavian and North American perspective. They also had to have a feel for the industry and the public dialogue on the topic. It is now up to our readers to evaluate their success. We gave our writers a free hand.

As the past has shown, predicting the future can often go wrong. While recognizing this, we encouraged our writers to accept the challenge. Without a doubt, this is the most dynamic section of the book. It was also the most difficult to write — and will probably be the easiest to criticize. We hope this chapter will inspire future development.

Our warmest thanks go to the Canadian-Finnish team who designed and wrote this book: Ken Drushka, Hannu Konttinen and Jorma Hinkka. We are also grateful to Kati Renvall, Timberjack Group's Communications Manager, who headed the project. Without Kati, 'Tracks in the Forest' would never have been ready in time for our 50th anniversary. With this book, we can mark our half-century in the industry with something more permanent than an anniversary celebration and self congratulations.

Having been in the field for a quarter of a century, I expected to find little new between these pages. This was not the case. Instead, I have gained a new perspective on my work — and an insight into how the industry appears to people on the outside, our writers included. So much the better, since you can never tell at the outset how such a project will turn out. This has been a rewarding project for us, and we certainly hope that 'Tracks in the Forest' will give you at least a few moments of reading pleasure.

Mikko Rysä
President & CEO, Timberjack Group

PLUSTECH
Timberjack
PLUSTECH

The latest development in forest machines, the Walking Machine, is equipped with a harvester head that fells, delimbs and bucks trees into programmed log lengths. The machine has three computers: one to control the motion of the walking legs, one to control the harvester head and one to control other machine functions, such as the hydrostatic drive.

Forest scarification in Finland in the 1930s.

A truck driver removes the chocks holding a three-meter diameter Douglas fir log before dumping it into a log pond at Tacoma, Washington, in the early 1950s.

# Introduction

Members of the human species are children of the forest. The evolution of civilization is intertwined with forests; at the center of our history lies the story of their use. Our ancestors believed that gods inhabit the forests, living in the trees. Remnants of these beliefs underlie the strong feelings, ranging from fear to reverence, many people still have about forests, and provoke intensely passionate debates over the way they are used.

Forests of one sort or another have existed for something like 300 million years, humans for perhaps one million. In our short span we have been both destroyers and creators of forests. Today, after several thousand years of forest use, eighty percent of the five-billion-hectare, pre-agricultural-era forest remains, covering about one-third of the earth's surface. Given the explosive growth of population during the past century, this gives each individual slightly less than three-quarters of a hectare of forest land.

There is enormous diversity in the world's forests. No two hectares are identical. There are broad types of forests, however, each type determined by the varied influences of geography and climate. Tropical forests occupy a broad area along the Equator, in Central and South America, Africa and Southeast Asia. Variations in these forests are caused by the amount of rainfall, which declines as the distance from the Equator increases. Tropical rain forests run along the Equator. Their most notable characteristic is the wide diversity of species they contain. The Amazon forest, for instance, hosts about 6,000 tree species; a typical hectare will have 85 to 90 species, with some areas containing 300 species per hectare. Relatively few of them are commercially valuable, and harvesting timber in these forests is both economically and ecologically challenging.

Farther from the Equator are the tropical seasonal forests, where rainfall is less continuous and comes at regular, annual intervals. There are far fewer tree species, mainly a mixture of deciduous and evergreens growing to heights of thirty meters or less. Valuable hardwood species, such as teak in Myanmar and Thailand, have been commercially harvested for up to 150 years. Extensive areas of these forests have been cleared for agriculture or the growing of cash crops.

As the distance from the Equator increases, and rainfall declines, the tropical seasonal forests merge into dry savannah forests. Most forests of this type are found in Africa. As rain levels drop even further, the savannah forests give way to dry, scrub woodlands consisting mostly of hardy bushes and shrubs.

In the Northern Hemisphere, at the greatest distance from the Equator, lie the boreal coniferous forests. They occupy two areas: a broad band across North America, from Alaska to Newfoundland; and across northern Europe and Asia, from Scandinavia to the Siberian Pacific. In this region there is a short, intense growing season, offset by long cold winters during which tree growth ceases. There are relatively few tree species, almost all coniferous, and large areas of land are covered with a single species, in part due to the frequency of intense fires which sweep through these forests. Various types of pines and spruces dominate in the boreal forest; birch and aspen are the major deciduous species.

The broad, mid-latitudinal area lying between the tropical and boreal zones contains a variety of different forest types. Perhaps the most important in human terms are the northern temperate deciduous forests found in the northeastern United States, western Europe, Korea and China. In Canada and Siberia, there is a narrow band of deciduous forest along the southern edge of the boreal forest. In Canada this is predominately an aspen-poplar forest, while in Siberia the aspen is mixed with birch. Compared to the tropics there are few species, the most prominent including the various oaks, maples, ash, beech, birch and elm. Through human history many of these species have been widely used, and their cultivation has, in many cases, become highly developed. Conifers also grow in these forests, but to a lesser degree.

Because of mild temperatures in the temperate zone, decomposition of forest litter is almost continuous, and a large portion of the nutrients are stored in the soil. Consequently, these soils are well suited to agriculture and extensive areas have been cleared for growing food crops. This factor has contributed to the concentrated growth of industrialized societies in temperate deciduous forest zones. In North America the deciduous forest blends into a temperate coniferous forest zone to the south. Around the Great Lakes and through to New England, this coniferous forest is composed of white and red pine, along with eastern hemlock.

Running in a great arc from Virginia, south to Florida and west through East Texas, this forest is dominated by pines, particularly longleaf and loblolly. The swamps and river bottom lands of the region are occupied by dense stands of gums, tupelo and other lowland hardwoods, interspersed with cypress.

Another version of the temperate coniferous forest is found on the Pacific Coast and in the western mountain zone. It is dominated variously by sitka spruce, Douglas fir, western hemlock, balsam, western red cedar and several pine species. These are all valuable commercial species which have been harvested for 150 years and longer. In the Southern Hemisphere there is a comparable temperate forest type, made up of quite different species. It is an evergreen forest consisting of pines, along with broadleaf evergreens such as eucalyptus. Relatively small areas of this forest grow in southern Chile, Argentina, southern Brazil, South Africa, Australia, Tasmania and New Zealand. Northern versions once grew in central China and southern Japan.

Another significant forest type is the Mediterranean. As indicated by its name, this is the once-productive forest which grew at various locations around the Mediterranean Sea, but is now 95-percent destroyed. A mixed evergreen forest on the lowlands, it changed to a coniferous forest in mountainous areas. A well-known example of the latter was the cedar forests of Lebanon. Other examples grow, or once grew, in parts of California, South Africa and Australia.

“

*We don't really know what hard work is since mechanization is like it is now. My crew, everyone of them's setting under air conditioning. Got as good a seat in the machine as you're sitting in right there. I went out there this year when it was 18 degrees and they was riding around with t-shirts on. Logging's not like it was. It's just not that kind of work any more. But it takes a lot smarter man to be able to run the kind of equipment we run today. Especially on the cut-to-length machine. On something like that, you gotta have a man that's really up on what he's doing, well trained, to be able to run cut-to-length machines.*

*From what I've seen all over the continent, it's no longer a bottom-end kind of life. Bottom-end job where you can't do anything else, you go logging. In the shortwood days — it's a terrible thing to say — if a man was real smart he didn't go out there and stack wood all day. He didn't cut it with a bowsaw and stack it all day. I've done it. I know. That wasn't very smart! But I've done it all day and it don't take a rocket scientist to be able to do that kind of work.*

*We've still got some people that cut logs in this country but we don't have near as many as we did. I'm talking about using the power saw to cut the tree down, trim it up, cut the tops out of it. We don't have that much of it anymore. The day of the real good sawhand is just about gone. All of them have just about got too old or died out.*

*Now, feller-bunchers cut all the trees, and then they're delimbed. Not all the time, but probably 50 to 60 percent of the timber we move in East Texas is moved strictly by mechanization. There's not that many sawhands any more that can directional fell a tree with a power saw. 'Cause people don't work like that any more. It's hard to get young people to come and do it. We've still got some, I'm not saying there's not any, but if I had to go hire a crew of sawhands tomorrow, I don't know where I'd start.*

TOM BURCH

Today, there is a new type of forest found in several parts of the world. Plantation forests are man-made forests which are beginning to play a significant role in the global forest economy. At present, a little more than 100 million hectares of plantations exist worldwide. About forty percent of them produce fuelwood, with most of the remainder grown for pulp wood and, to a lesser extent, lumber. With few exceptions, plantations are grown for commercial use and supply an increasing share of global timber production.

Human beings have always depended on forests. Initially, we used them as places to live. We hunted in them for game, foraged for fruits and nuts and gathered fuel. Our relationship with our habitat was essentially no different than that of any other animal. The development of settled agricultural economies to replace those based on hunting and gathering required the clearing of forests. As settlements grew into cities, the need for agricultural lands and fuelwood increased. In some parts of the world, the pre-industrial use of forests for fuel or farm land was extensive enough to deplete or destroy them. This occurred in most regions of the Mediterranean. In other places, laws were passed to control deforestation. As early as 300 BC, Egypt controlled timber harvesting and practiced reforestation in the Nile Valley. Much of this forest survived until the eleventh or twelfth centuries; the last stands disappeared only at the beginning of the present century.

Humans did not begin using forest trees extensively until about the sixteenth century. Generally speaking, the means were not available before then to convert logs into useable boards and timbers. Exceptions were the Pacific Northwest of America and Japan, where easily split tree species were used for building houses and other purposes. What precipitated the heavy use of timber was the development of navigational skills permitting European voyages of discovery across the world's oceans. Just as the forests of Lebanon had provided timber for a succession of seafaring Mediterranean civilizations, the forests of Europe provided materials for the building of ships to sail the Atlantic, Pacific and Indian Oceans. These explorations led to voyages of colonization and the growth of large overseas empires. With national power dependant upon possession of vast fleets of cargo ships, and the naval forces needed to protect them, timber became the most strategically important material in the world. A global timber industry was created, consisting of logging and milling components. As a by-product, large amounts of lower grade lumber were produced which found a ready market in the construction of buildings, bridges and other structures. Once established, this market developed its own momentum and put its own demands on forests. The advent of the Industrial Revolution in the late eighteenth century created enormous demands on forests. Vast amounts of wood were consumed as fuel or in the form of timber used as mine props and to build railways. Railway steam engines started large numbers of fires which burned huge areas of forest.

In the nineteenth century, the growing demand for timber spawned a distinct new industry, logging, to bring trees from the forest to mills for processing into useful products. In parts of the world, particularly Europe and Japan where forest exploitation had existed for centuries, timber-extraction practices evolved in tandem with the development of silviculture which sought to replenish forests. In more recently colonized areas, particularly North America, growth of the logging industry coincided with a mechanical revolution precipitated by the invention of steam engines. For almost a cen-

“

*A man of my age, when I think about the old cable skidders and stuff, I don't call that mechanization. I know it was, but what I really mean by mechanization is saw-head machines, grapple skidders, hydraulic loaders — stuff that really took the work completely out of logging is what I think of as mechanization. A man's hands never touch the timber. There's no manual labor, when I think of mechanization. A man, he sits up here running the machine, punching buttons. He gets that tree on the truck and hauled to the mill without any sweat. I call that mechanization.*

Tom Burch

tury, until after World War Two, most specialized logging machinery was developed in North America, where the primary task was the liquidation or exploitation of forests. The diversity of forests and logging conditions on the continent led to a proliferation of diverse logging technology. When European logging methods were eventually mechanized, the technological developments were more heavily influenced by a need to manage forests carefully. In the past decade or two there has been a synthesis of these two strands of logging technology, producing a range of modern logging equipment used around the world.

Today, we are in the early stages of developing the machinery needed to utilize forests in a manner in keeping with our evolving perception of their role in our lives and our civilization. New machines are not so much logging machines — although that is one of their primary functions — as forest machines. They are designed to operate without inflicting permanent damage on the forest. They are built in recognition that the forest is a habitat for humans and other creatures, a complex entity from which we must extract certain essential materials without damaging the ability of the forest to maintain its productivity and diversity. And they are also built to harvest timber efficiently within the economic constraints of competing methods.

Modern forest machines include two new, essential components. The first is a range of computers which can be programmed to make certain decisions or choices, while operators concentrate on problems which have as much to do with silviculture as with the task of extracting logs. The second component is an attitude or frame of mind on the part of the operator. The new machines make it possible, in fact make it mandatory, for the modern logger to focus attention on the forest he will leave behind after his job is finished.

“

*I'll give you an example of what logging evolution did to a logger in Tennessee. He was an ex-West Virginia coal miner. He lost a leg in a coal mine accident so he decided he wanted to be a logger. He bought an old truck, a couple of chain saws, and a little D3 Caterpillar — a real slow operation. He also had a winch, with a block up in the tree and powered with an old Chevrolet engine. They would drag the trees into his little landing, his boys would cut them into log lengths. They would each get on the end of a log and put the tongs in, or hook into each end of the log. The old man'd start the engine, put it in gear, and winch one log at a time onto a truck. Now that was pretty damn slow.*

*So I went down and I said, "Harry, what you need here is an automatic loader. You need a Prentice loader and you can pick up one, two, three, four logs at a time and put them on your truck."*

*"Well, show me how it works," he said. So we had a little demonstration for him. "This is a great idea," he said. And he bought it. However, what he didn't realize was that now the loader sat and waited. There was not enough wood being brought in for him to load.*

*"Now what you need to balance this system with is a log skidder," I told him. "I'll sell you a Timberjack log skidder, a real nice skidder, and you can bring more trees out, ten trees, eight trees at a time, cut them up and put them on your truck." So he bought the log skidder. Now he only had one truck. He didn't have enough trucking capacity to balance his system. So he had to buy two more trucks, and when I was finished with the guy he had a real nice little system. One man and two sons system. Good productivity.*

*Another example in that same area of Tennessee — a man and his boy were loading long hardwood, cutting it up at the truck and cross-hauling it. They put two skids on the side of the truck. They had a little tractor on the other side with a cable that would drag the logs up onto the truck.*

*"How long does it take you to load this truck?" I asked them.*

*"Oh, we make one load a day."*

*"One load a day!" I said. "Well, that means you two guys here, you and your son, are loading one truckload in one day. You making any money?"*

*"Oh yeah, we make enough to live on."*

*"Just wait here one minute," I told them. "Don't finish loading this truck." We had the first knuckleboom loader in the region just half a mile up the mountain, so I went up and got him. He came down and finished loading the truck in five minutes. That opened the guy's eyes. "Christ," he said. "This is great!" And he bought a loader. In that one valley — Shady Valley, Tennessee — I sold twelve loaders within a year and a half. People saw one, next guy bought one, next guy. It was exciting times.*

Mal Crawford

The significance of these new machines for the forest industry, and for human society, should not be underestimated. The history of machine use in the forests is not entirely laudable, at least by today's standards. The most widely-used logging system to emerge from the steam era was probably high-lead cable yarding. For the forest, it was the most destructive form of logging devised to that time. Similarly, the rubber-tired skidder, in its early manifestations, was used by some loggers to inflict enormous damage on forests. It is only necessary to look at advertisements for these early machines. Typically, they are shown buried to their axles in mud, skidding logs out of swamps. Or they are pictured heading down a steep slope, a huge turn of logs on behind, digging a trench that will funnel silt and topsoil into the nearest stream at the first rain. The people who made and used these machines wanted them tough enough and versatile enough to power their way through the most adverse conditions. Swamps, muskeg and steep slopes were seen as challenges, difficulties to overcome. Only later did it become evident that these efficient machines were damaging the very forests upon which their existence depended.

The conditions that created or permitted such an exploitative attitude toward the forest have changed. Today the public demands, foresters understand and most loggers and machine makers know that their livelihoods, along with the well-being of civilization, depend upon the ability to utilize the forests without harming them. Some find it difficult to give up the old ways. Others, the new generation of loggers who seek the best ways for their children and grandchildren to earn their livelihoods from the forest, have a different attitude. Try extolling the superiority of the old ways to a modern Alabama pulp logger who spends his days working in air conditioned comfort, listening to music on his tape deck while he carefully thins a stand of loblolly pine, leaving behind an intact, productive forest.

For centuries people have gone into the forest — for fuel, for food, for timber, to fill some basic human need. Perhaps what has motivated them most is a love of adventure. That is certainly a part of what attracted the hunter, the axeman, the logger. Man's use of machines in the forest has been part of the adventure. Today that adventure takes on a new dimension. Today we understand that we have become who we are through our use of forests, and that our fate hinges upon how we use forests in the future. The adventure of men and machines in the forest has only just begun.

“

*I had read an article Alan Buell wrote on stabilization of woods labor, which really appealed to me because having grown up in the woods and seeing all these people who had no home, living in bunkhouses — what a rotten way to make a living. As a young guy you have these feelings that you're going to change the world. Two things appealed to me: logging mechanization and the stabilization of woods labor. When I read this article I thought I'd really like to meet the guy. So I went to Montreal to the CPPA during my last year and interviewed him — he was the manager of Longlac Pulp and Paper. I was so impressed with his manner and his thinking that, notwithstanding the fact that they didn't have a lot of mechanical operations, I went and I made a deal to work with them. But in logging, not in forestry. So I went there and I was what they called an assistant camp foreman. That was '52, and in '54 one of the foremen got kicked by a horse and broke his leg so I got a chance to be chief foreman. And that was my start as a camp foreman.*

*The key to mechanization was stabilization of the work force, because it's pretty hard to mechanize if you've got transient labor. They went hand in hand. And the reason I went to work for those guys was the fact that this guy had a social conscience as well. Not only a social conscience, but it was a social conscience that was going to make money for the company. A different view, a different approach. It really appealed to me, having been brought up in a logging camp and seen the people come by the door every spring and go back out every fall.*

LEON POND

Two handfallers carefully drop a big Douglas fir tree into standing timber on the British Columbia coast in 1919.

# From Muscles to Machines

For thousands of years, humans have used their own muscle power to bring logs out of the forest for use as fuel, lumber, paper and other necessities. To avoid this drudgery, they began devising easier, more efficient ways of falling trees and transporting them to the mills. The great diversity of physical conditions found in the world's forests means that for any situation, there is sure to be a better way of doing the job, a labor-saving machine suited to the task.

Since water is the most readily available medium for transporting timber, the timber trade evolved in close proximity to rivers, lakes and oceans. Water powered the earliest sawmills, and the best, most accessible stands of timber were found along the rivers and streams on which the mills were built. From the outset, men exerted their muscles and devised ways of getting logs into the water where they could be floated to the mill. Until the invention of the first logging machines during the last half of the nineteenth century, the task was approached in essentially the same manner the world over. Although systematic commercial exploitation of the Lebanese cedar forests is known to have been well underway by the eleventh century BC, little knowledge has survived concerning the logging methods. What can be deduced is that no machines were used in this work and, as is most likely, the methods employed were similar to those used for most of the next 3,000 years. By the beginning of the Christian era, forest management was being practiced in Greece and parts of the Roman empire. Again, the methods used to harvest timber, are not clear. The earliest records indicate an almost-universal use of controlled burning to fall and buck trees, and human labor to move them directly to where they were used, or to a nearby body of water.

For example, during the twelfth century AD, a massive timber harvesting enterprise was launched in the rugged Qin Feng mountains of China to obtain materials for the reconstruction of the city of Kaifeng. A road was built, including a five-kilometer-long bridge, over which timber was transported in some type of wheeled vehicle. In the late-eighteenth century a logging operation in southern China used ropes to yard logs to a river with capstan devices turned by oxen, mules or teams of men. Flumes and slipways several kilometers long were also built to move logs to the river, where they were floated another 50 kilometers to the sawmill.

In Japan's Kiso Forest, during the 1850s, loggers used axes with long, narrow heads to chop deep holes into standing trees. Fires built in these holes felled the trees, which were then hewn into squared timbers and lowered down steep slopes with ropes snubbed around stumps to an elaborate network of flumes leading to the river below. There the timbers were boomed into rafts and floated to log storage ponds near the mill towns. At approximately the same time, in Australia, men were employed to carry flitches, slabs and bundles of shingles out of the forest on their back. Others rolled and levered logs along skids laid in cleared paths. In Tasmania, convict "centipedes" carried large sawlogs out of the forest as a form of punishment.

In the 1880s, along the steep-sloped inlets of British Columbia and Alaska, handloggers felled Douglas fir and Sitka Spruce trees up to two meters through toward the water onto a grid of saplings. With luck, the trees skidded themselves into the water, where they were topped and limbed, boomed into rafts and kedged or towed to the nearest mill. If they stopped before they reached the water, loggers used powerful hand jacks — locally made Gilchrists and Ellingsons were among the first logging tools manufactured — to raise up the leading end of the log. Another jack was placed at the back end of the log to nudge it ahead over a bed of peeled saplings. With luck, on a downhill slope, the log ran to the water; on level ground, it moved only a meter or two. Handloggers used this method to move 100-ton logs one hundred meters or more across level ground to water.

In the Appalachian Mountains of the eastern United States, until well into the twentieth century, a handlogging technique known as "ball hooting" was employed to roll heavy hardwood sawlogs

> I started my work in the woods at the age of 16. I was in the woods from 1896 to 1906. When a young fellow started in the woods they usually put him at some light job such as sleigh hauling and road cleaning. The person who did the road cleaning was known as the "road monkey." The road monkey would sweep the ruts and repair them. The horses wore sharp shoes and these would cut into the ice, thus causing chips of ice to fly into the ruts. Also, horse droppings would fall into the ruts and it was the road monkey's job to sweep the ruts clean.
>
> George Eitel

down steep slopes. Teams of men, armed with peaveys, rolled the logs along trails cut through the forest and downhill to a landing where they were loaded on wagons or trucks for transport to the mill. As late as the 1930s, along the north shore of Quebec's St. Lawrence River, a hand-logging system used for centuries in the mountain regions of Central Europe was still employed. Known as "chienning" (dog's work), it involved the use of small hand sleighs to haul pulpwood bolts downhill for distances of up to 300 meters. Snubbing ropes eased the sleighs down steeper slopes. As late as 1970, a method known as "kuda-kuda" (horse-horse) was used to log in the peat forests of southwestern Sarawack. Bucked logs were rolled by hand over the slash on saplings to skidroads built from more saplings, where they were loaded on small wooden sleds and hauled by five or six men, one log at a time, to the nearest railway line.

Pre-mechanized commercial logging began in British North America early in the eighteenth century, primarily to provide spars for Great Britain's Royal Navy. After the American Revolution began in 1776, most of the British timber trade shifted north to the extensive white pine stands in Nova Scotia, New Brunswick and the Ottawa River Valley. A smaller, domestic US trade established the beginnings of a forest industry in New England.

At that time, the major source of European timber was the Baltic region. In 1806, when Napoleon blockaded the Baltic to British ships, attention focused on Canada, and the timber business there underwent an immediate, enormous expansion. The logging methods employed were patterned on those which had evolved in the forests surrounding the Baltic Sea, where throughout much of the eighteenth and most of the nineteenth centuries, hundreds of thousands of men and tens of thousands of horses earned their living in the forests. Trees were felled manually, by axe and, later, by saw. The work was hard, and as time passed, it took place at increasingly longer distances from settlements. Wages were low compared to that paid to factory workers in the cities. Horses were easily exhausted pulling oversized loads in the freezing weather. At night men crowded into cramped quarters with dirt floors. Sometimes a whole winter passed before a family man could make it home. Rye bread and potatoes were the staples; lard and Baltic herring an extravagance. Nothing was heard about vitamins, pensions or work safety. Hard work would have worn out a man by the time he reached the age forty. During winter, trees were dragged by horse and sleigh to the sides of the lakes and rivers. When there was little snow, the bulk of the timber remained in the forest.

During summer, timber was floated to sawmills built along the waterways. A nothern logger of this era was like a cowboy of the north, transient and free to wander. Forest workers in the settled areas farther south were mainly cottagers and tenant farmers. Even after the industrial revolution, steamships were rarely seen during timber drives, and only a few loggers ever saw a railway; they were for the gentry.

By the 1830s, the population of North America, particularly the United States, was increasing rapidly. As eastern cities expanded and the movement west began, the timber industry followed behind. During this period, there was a rapid development of logging through the Great Lake states of New York, Michigan and Wisconsin, and into Minnesota. In these northern forests, reaching up into Ontario and Quebec, with their long, cold winters and heavy snowfall, a system of logging evolved based on the extensive use of horse power. It was a system which took advantage of the country's harsh climatic conditions. Prior to this time, the skidding of logs was done with oxen during the summer. Oxen were less expensive than horses, worked more efficiently on soft or wet ground, and could always be eaten if they broke a leg and had to be killed. Nineteenth-century logging in the north woods of America, however, became essentially a winter occupation, with most of the crews recruited from farms which were all but inoperable during the winter months. After the first snowfall, loggers loaded horse-drawn sleds with supplies and travelled up the river valleys to remote camps where they stayed until spring, felling and delimbing trees with axes and bucking them into logs with the inefficient hand saws of the era. In Canada, until the 1860s, most logs were hewn into squared timbers and exported to Europe through the port of Montreal. Later, the bulk of the harvest was sold to US buyers as logs, or as lumber cut in Canadian mills.

Logs were skidded to a haul road with a team of horses, then cross-hauled or parbuckled onto large sleighs. Sleighs were pulled along carefully prepared ice roads to the nearest river, where logs were decked until the spring breakup. At this point, river-driving crews took over. During the early stages of this era, logs were hand hewn into square cants which were formed into large rafts and floated down the river. Unsquared logs were simply dumped into the freshet-swollen water and herded downstream by rivermen.

The hauling phase of this method evolved into a sophisticated system of logging centred around the use of horses. A great deal of attention focused on the design of sleighs and equipment used to prepare the haul roads over which they ran. A network of main roads, with secondary branch lines, was prepared in the summer and fall, with stumps and rocks cleared and depressions filled to create a level surface. After the first major snowfall, horse-drawn rollers constructed from saplings were dragged along the roads to pack the snow. Then sleighs carrying large, heated tanks sprinkled water on the snow, building up a layer of ice. A specially designed plough was pulled along the road, cutting ruts into the ice to guide the sleigh runners. Sand or hay was stockpiled at steep downhill grades to use for braking the sleighs. Steel-shod runners, hauled over a well-maintained ice surface, met little resistance, and the system made efficient use of the limited power available from a team of horses. Loads of one hundred cubic meters, hauled up to four miles to the river, were common.

Elsewhere, the use of animals for logging evolved in directions determined by local conditions. In general, there was an initial use of oxen, which were then abandoned in favor of horses. In warmer climates, where ice roads were not feasible, other methods were invented. In Australia, for instance, a variety of techniques were devised for "snigging" or skidding logs to a landing or dump. Horses and bullock were used to drag logs down chutes gouged in the ground or constructed with parallel logs. Logs were loaded on jinkers, a set of wheels and an axle, both made of wood, by the simple expedient of digging a hole for the wheels to drop into, then rolling the log onto the axle and chaining it down. In some situations, one jinker was used, with the rear of the log trailing on the ground; in others, each end of the log was loaded on a jinker, and towed like a wagon. Similar devices, called "go-devils," were used in the southern US, where mules were preferred over horses because of their superior ability to work in hot weather.

Also in Australia, as in New Zealand and parts of the American south, horses were used on tramways. Narrow gauge rail lines — often using saplings or steel-clad timbers as rails — ran up gently sloping grades into the logging site. Horses pulled small wooden carriages with flanged wheels along the tram lines, where logs skidded out of the woods were loaded from ramps, or by cross hauling up a pair of skid logs. Snubbing lines eased the carts down steep grades. In Japan, during the late eighteenth and early nineteenth centuries, loggers rode loaded tram cars down steep slopes, controlling their speed by pulling on chains attached to wooden brake blocks. A variation of the tramway was developed during the late nineteenth century in the US Appalachian Mountains. A wooden capstan turned by oxen or mules hauled in a Manila rope which pulled a carriage loaded with logs along a set of wooden rails.

During the late 19th century, in various parts of the US, another system of horse-powered logging technology evolved which was far more mechanically advanced than sleighs and trams. Most of this development took place in the warmer southwestern pine forests, where a combination of big logs and dry conditions prevailed. It evolved from the use of two-wheeled carts which were loaded with logs in two ways: on top of the axle, or slung beneath it. When logs were on top, the cart was known as a bummer. Various versions of the bummer evolved and carried through into the mechanical era. In the horse-drawn period, advances in bummer design focused on self-loading devices. One of the most popular was designed by John Lindsey, a highly talented wagon maker from Laurel, Mississippi. His loading system consisted of a stiff tongue which extended a short distance over the rear of the axle, with a set of log tongs attached to it. The bummer was backed up against the log with the front of the tongue raised vertically. Tongs were attached to the log and when the horse team pulled on the tongue the front end lowered, raising the log, attached to the rear end onto one of the bummer wheels. As the bummer moved ahead, the wheel rolled the log up onto the axle for skidding to the desired location.

Bummers saw limited use because their wheels had to be small to facilitate loading. Simple mechanics dictated that larger diameter wheels supported larger loads and easily rolled more over stumps and rocks, so a device known as the "big wheel" was developed to haul big logs. It carried logs slung underneath the axle and in order to carry logs two meters in diameter, wheels of four meters or larger were required. Sylas Overpack of Manistee, Michigan, designed one version of the big wheel using a stiff tongue featuring three-meter diameter wheels. Logs were loaded on the wheel by backing it astride them. The tongue, similar in design to the Lindsey bummer, was raised to the vertical position and the logs chained to its rear end. When the team pulled forward, the tongue was pulled to a horizontal position, raising the logs, which were then chained to the axle.

There was no braking mechanism on the stiff-tongue wheel; on steep slopes they had an unfortunate habit of overtaking the horses pulling them. To overcome this problem, Joseph Carroll of Beaumont, Texas, developed a sliding tongue. Known as a slip tongue, it raised the logs by use of chains on a drum activated by the forward sliding motion of the tongue. On a downhill grade, if the wheel gained on the horses, the tongue slipped back, lowering the logs and braking the wheel. The teamsters rode on an English-style saddle on one of the horses, or on the tongue. It was a colorful, often hazardous occupation, with the teams running at full gallop, followed by wheels towering to twice their height, with huge logs slung underneath, dust rising in clouds and the shouts of teamsters urging their horses on to keep ahead of the thundering wheels.

During the late nineteenth and early twentieth centuries, hundreds of these big wheels were built, most of them by wagon-wheel manufacturers. One such company, the Stockton Wheel Company, began making big wheels in the 1880s as part of its line of logging wagon wheels. This pioneer logging equipment company, started by the Holt brothers in New Hampshire during the first half of the century, moved to California in 1860. The firm eventually became a component of the Caterpillar Tractor Company, one of the major players in the mechanization of logging. The earliest use of crawler tractors in logging was to pull big wheels in the California sugar pine forests.

During the years in which big wheels were evolving, two other developments took place which eventually eclipsed the use of horses and other forms of animal-powered logging systems. Both innovations — railways and wire rope logging — were based on the use of steam-powered machinery.

Railways came first. During the reconstruction period following the American Civil War (1861–65), a network of rail lines spread across the North American continent. In many parts of the United States, rail companies received grants of large tracts of public forest as incentives to build their lines. Railways provided access to timber stands unreachable by river-driving methods and precipitated a general movement westward and southward by the lumber industry.

Typically, branch logging lines ran off mainline railways, up river valleys and their tributaries. Often these were narrow-gauge lines, using smaller equipment, occasionally consisting of trams pulled by horses or small steam engines. The full development of railway logging was delayed by a lack of suitable equipment. In the early stages two types of locomotives were readily available — large, mainline engines and much smaller engines used in construction and other industrial applications. Neither was particularly suitable for logging. The mainline engines were too large, requiring expensive, heavy-gauge tracks and substantial roadbeds. They also were difficult to maneuver in the tight confines where loggers worked. Small engines, with their rod-driven wheels, lacked the power needed to ascend the steep grades encountered by loggers.

In the mid-1870s, a Michigan logger, Ephraim Shay, began building a new type of locomotive for use by loggers in his region. A compact, powerful machine, it transmitted power to its drive wheels from vertical steam cylinders through a driveshaft and set of bevel gears. After selling his idea to the Lima Locomotive Works in Ohio, Shay went back to logging. By 1893, 450 of his engines had been sold. Its powerful steam engines and unique driving system enabled the Shay to operate on hills two to three times as steep as those accessible to conventional locomotives. The driving system was also much easier on the light track used in logging operations.

Before the turn of the century, at least two other companies began producing geared locomotives designed for logging. One was the Climax, manufactured in Pennsylvania. It had an engine mounted on each side of its boiler to drive a crosswise driveshaft connected to the wheels by bevel gears and a longitudinal shaft. About a thousand Climaxes were manufactured before production ceased in 1928. The Heisler locomotive, introduced in Erie, Pennsylvania, in 1894, used a different configuration of engines, a central driveshaft and bevel gears to power its drive wheels. It was the least popular of the locomotives designed specifically for logging, but it remained in production until 1945.

The application of railways to logging was an adaptation of existing technology. Cable logging, on the other hand, was the first significant mechanized technology designed specifically for the harvesting of timber. It developed primarily in two regions, the American south and the Pacific Coast, where conditions required the skidding of large logs over wet or swampy ground that did not freeze enough to permit winter hauling. Cable logging began in California in 1882 when John Dolbeer, a partner in a Eureka lumber company, patented a steam-powered winch with two horizontal drums. The following year, he brought out what he called an "Improved Logging Engine," which had a single, vertical gypsy head, or capstan, on it. This device, probably copied from steam capstans used on sailing ships, revolutionized West Coast logging.

Dolbeer's yarding machine, and its successors, became universally known as a "donkey." A line — initially Manila, then wire rope — was hauled out from the donkey by a horse and attached to a log. The spool tender took a few wraps around the revolving gypsy head, pulled it tight and reeled in the line and log. Soon, an additional drum and line were added to pull the main line back out. Larger drums were built to facilitate greater reach. A version of this machine, known as a "roader," pulled logs along an established skid road, which might be fitted with crossties of the sort used by teamsters, or fore and aft skidways. Some years after the turn of the century, inventive loggers thought of hanging a block from a convenient tree and threading the mainline through it out to the logs. This lifted the leading end of the log off the ground, over stumps and other ob-

"

*Finland's first mechanized logging site was at Koilliskaira in 1912. Kemi Oy's forest manager, Hugo Sandberg, had ordered two Lombard steam-engines from the Phoenix Manufacturing Co. The logging site was a big one; with hundreds of men and horses. Because of the inadequate and unvaried nourishment, the lumbermen got scurvy.*

*Sandberg had read in Jack London's books that horseradish was a good remedy for the disease. So he bought all the horseradish that was available, but the lumberjacks wouldn't touch it because of its taste. Then two cow-sheds were built to get fresh milk. But to ensure continuous milk production, a bull was also needed. The bull was transported by sleigh from one shed to the other as needed. These trips were so much to the bull's liking that when it heard the sound of sleigh-bells, it started to bellow. When it was let loose, it ran to the sleigh by itself and threw itself in on its back. The lumberjacks watched the bull's trips with envy: there were no milkmaids tending the cows.*

Esko T. Helin

structions, and was known as "high lead" logging. As donkeys became larger and more powerful, high lead logging became a brutally efficient skidding method. The yarding of big timber by powerful steam yarders destroyed every other tree on the site. In this manner, the kind of clear-cut logging that would one day give the industry such a bad name was born.

The rapid acceptance of Dolbeer's donkey spawned an entire industry of logging equipment manufacturers. Within twenty years or so, every city of any size on the West Coast had a company producing steam-logging machines — Washington Iron Works in Seattle, Willamette Iron & Steel Works in Portland, Empire Manufacturing in Vancouver, and Westminster Iron Works in New Westminster, BC, to name only a few. By 1908, about 4,000 steam donkeys were at work on the West Coast.

At approximately the same time, a number of manufacturers appeared in the eastern US, the most important being Clyde Iron Works of Duluth, Minnesota, and Lidgerwood Manufacturing of New York. The latter entered the steam skidder business in cooperation with Horace Butters, a Michigan logger who in 1883 designed and built a cable-logging system utilizing standing trees to yard logs from the woods and load them on railway cars. The system used a steam-powered hoisting winch manufactured by Lidgerwood and was built to yard logs out of swamps and pot holes in the Great Lake forests. It consisted of a steel tower mounted on a railway car and a set of winches to operate a skyline, a main line, and a haulback line to pull a carriage out along the skyline. An additional line and winch enabled the carriage to be pulled toward the machine independently of the main line, providing slack on the main line to facilitate the attachment of logs. Initially, this machine was rigged with Manila ropes, severely limiting its capabilities, but the coincidental development of wire rope enabled the rapid development of the Lidgerwood machine in particular, and cable logging equipment generally. Aware of the impending depletion of northern timber, Butters moved south and obtained extensive holdings in the Green Swamp of North Carolina. There, he and Lidgerwood's engineer, Spencer Miller, developed a full-blown version of the cable-logging systems slowly evolving on the West Coast. At about the same time, a similar line of equipment was being developed by J.R. McGiffert at the Clyde Iron Works in Minnesota.

Over a span of about four decades, until the onset of the Great Depression of the 1930s, two distinct streams of steam-powered cable-logging systems evolved in North America. The West Coast machines developed into a wide range of skidding donkeys, reflecting the huge size of the timber and the rugged logging conditions encountered there. These machines were normally mounted on log skids, which could be pulled onto rail cars and hauled to the logging site where they were set beside the tracks and rigged through wooden spar trees as much as seventy meters in height. If these trees were conveniently growing on site, they were trimmed, topped and guyed for the task; if not, they were moved to the site and raised by a rigging donkey.

There were many possible rigging configurations, ranging from a high-lead setup for skidding over distances of up to 250 meters, to a wide range of skyline systems which reached out several hundred meters. High-lead systems dragged the logs over the ground, with the leading end off the ground; skyline systems lifted logs entirely off the ground, carrying

> "
> *I reckon the first thing I remember about the woods was back about 1907. A terrible storm hit the Alabama and Mississippi coasts and tore up timber everywhere. Me and my brother pulled on the same handle [of a crosscut saw] as we helped our daddy work up some of that stuff.*
>
> *I helped log this same tract back in forty-one or forty-two. We just picked it over then. Sort'a cut what we wanted to and what we could get with what we had to get it with.*
>
> John Leslie King

them over ravines, ridges, stumps and other obstacles. In some cases a second skidder — and, occasionally, a third and fourth — dragged itself out to the back end of the setting to log an area beyond the reach of the "home" spar. Logs were piled, or "decked", at these outlying spars and later skidded or "swung" to the home spar in an operation known as "cold decking."

At the landing, faster means of loading onto rail cars had to be devised to deal with the rapid accumulation of logs. On the West Coast, the preferred loading systems employed a variety of devices utilizing the spar tree and the skidding machine or, usually, a separate loading donkey. Blocks were hung from guylines, positioned so that tong lines could pick up logs and lower them onto rail cars. A refinement of this system using a specially built donkey, called "duplex" loading, was a fast and much-favored system at hiball logging shows. Other methods employed a boom or rack mounted on the spar, with one or two tong lines hung from it, and additional lines used to swing the boom over the rail car. The former utilized a method known as "heel booming" in which a log was caught off-center with the tongs, raising the shorter end up against the heel of the boom or rack, then swinging it over the car and lowering it into position. A Vancouver-area logger, Claude McLean, designed a set of steel fittings to simplify construction of these booms, which were highly popular from California to Alaska until the 1960s, when the use of wooden spar trees ceased.

In eastern North America and the southern US, a different set of cable machines evolved. Although smaller counterparts of the West Coast donkeys were used in these regions, portable skidders mounted on rail cars were more popular. These machines were used in various parts of the world in the early years of the century. In 1910, for example, the Southern Brazil Lumber Company imported four Clyde skidders and two loaders for use at its operations on the Rio Negro River. These machines and a short-line railway delivered more than one thousand cubic meters of pine and imbuyah a day to the company's mill at Tres Barras. A number of cable systems were also used in Ontario shortly after the turn of the century, most of them Lidgerwoods.

Another type of portable skidder evolved for logging the timber-rich cypress swamps found throughout the American south from Virginia to East Texas. The first cable yarders used in the swamps were called pull-boats. They were barges on which a steam engine was mounted, powering two large drums, one with a main line up to 1,500 meters long, and a second, haulback drum. The pull-boat was towed up a main channel to the logging site where a circle of radiating skidding channels were dredged through the swamp. Trees were felled toward the channels where lines were attached and the logs were skidded — actually, floated — to the pull boat, where they were boomed up and towed to the mill or rail head. An early version of the Lidgerwood overhead cable system with a spar mounted on the barge was tried at a few locations in the swamps, but it was not successful.

A New Orleans logger, William Baptist, developed an improved pull-boat system, later acquired by Lidgerwood, utilizing two drums and an endless cable setup, with a large block at the tail end. In this system, when one side of the line was pulling in logs, the other side was taking empty tongs or chokers back to the setting. A refined version of this setup included a high-speed haulback drum to speed up the skidding cycle.

In the early 1890s, Lidgerwood developed a swamp-logging skidder mounted on a rail car, which ran along lines built deep into the swamps. Some of these tracks were precarious propositions, supported by cross logs and pilings driven into the mud. Lightweight "dummy" locomotives were used on spur lines to haul the loaded cars to the mainline, where bigger locomotives took over. This approach soon dominated the swamp-logging business, leading to the evolution of a range of portable cable skidders — called "snakers" — used throughout the southern pine forests and, to a lesser extent, in the northern forests.

The ultimate refinement of these machines was unquestionably the Lidgerwood Portable Logger. It was a huge machine, mounted on a 30-meter rail car, with numerous winches to control a multitude of lines: a skyline, up to four mainlines, a haulback line, loading lines, and a line for spotting rail cars. A 30-meter steel tower was mounted at one end of the car and at the other a steel loading boom. On location, the skidder platform was raised on steam jacks resting on the railway ties, and its car removed to allow log cars to be pulled underneath it with the spotting line. In operation, a string of empty cars was spotted behind the skidder and successively loaded as they were pulled under the platform. A crew of 19 or 20 men operated the skidder. The Clyde company developed a similar machine, but its trucks folded up underneath the skidding platform to facilitate passage of rail cars underneath it.

The Clyde company's greatest contribution to logging mechanization in this era was probably the McGiffert loader, named after its design engineer. Aside from the West Coast, North American cable yarders favored the use of rail-mounted, mobile loading machines. With smaller logs to handle, it was more practical to use a portable loader than in the West. The McGiffert loader, the Cadillac of steam loaders during its day, was a heel-boom type mounted on powered wheels which swung up on hinged supports, leaving the machine standing on four curved legs resting on the rail ties. Empty cars were pulled beneath the loader, and loaded in front of it with a fixed boom. The loading line was equipped with a slack-pulling device, which enabled the skidding of logs over short distances. With a nine-man crew, these loaders were capable of loading 2,500 cubic meters of southern pine logs in a day.

A variety of other mobile loaders also were used throughout the south and parts of the north and northeast. The Barnhart loader, developed in Pennsylvania, was the first of another type. It rested on a set of portable tracks laid on the rail cars it was to load and pulled itself along the tracks with a wire rope. The engine, boom and winches were mounted on a turntable with a 360 degree swing, the first use of this technology. The Marion and American companies produced similar machines, with the added improvement of gear-driven turntables.

The advent of steam-powered logging equipment launched an unprecedented assault on the southern pine and cypress swamp forests of the United States. Between 1910 and 1927, more than five million cubic meters of cypress were taken from the Okefenokee Swamp alone. By the early 1930s, about 24 million hectares of longleaf pine forest between the Atlantic and East Texas had been clear-cut, producing more than two billion cubic meters of logs.

The impact of mechanized steam power was scarcely felt in the forests of the northeastern US and eastern Canada during these early stages of mechanization. With two exceptions, there was little development of logging machinery specific to these areas. The first exception was a steam warping tug known as an "alligator," invented and built by John West in Simcoe, Ontario. It consisted of a heavily reinforced scow-shaped hull, about twelve meters long, on which sat a powerful steam engine, equipped with a Waterous boiler, and a winch strung with 1,500 meters of heavy cable. A large-fluked warping anchor was attached to the end. The tug was powered with paddle wheels and, in later stages, propellers. The alligator's function was to assist in the driving of logs down northern rivers and across lakes. It collected logs in a bag boom at the mouth of a stream and kedged or warped them to the lake's outlet with its winch and anchor. Once the logs were driven through the rapids, or flumed down to the next lake, the alligator used its winch line to portage itself overland to the next lake. Only 207 alligators were built between 1889 and 1934, when production ceased. Most of them were used in Ontario and Quebec, although some were used in the northeastern US and a few in southern swamps.

The most significant steam machine to evolve in the northern woods was the steam hauler. In the mid-1880s, loggers in Wisconsin and Minnesota were beginning to encounter ice-road hauls of more than eight kilometers, the practical limit for two trips a day with a team of horses. Various loggers, aided by local blacksmiths and amateur mechanics, began converting steam threshing machines into log haulers to pull sleigh-trains of logs. Initially, these machines met with limited success as their steel, lugged wheels chewed up the ice roads, which had to be maintained in good condition to allow easy movement of the sleigh runners. This problem was solved in 1900 by a self-taught, Maine millwright, Alvin Lombard, who designed a unique steam hauler for a Kennebec logging company. In profile, it resembled a locomotive, with a horizontal boiler, an underslung engine and cab behind. The drive mechanism was what set the Lombard machine apart from any other, and eventually revolutionized the development of mobile industrial equipment. The machine's twin steam engines powered two sets of sprockets which ran inside two sets of endless steel tracks equipped with steel pads. As the sprockets moved themselves along the inside of the tracks the latter were lifted at the rear, carried forward over the tops of the sprockets and laid down on the ground ahead of the forward sprocket. Lombard's early models were steered by a bobsled arrangement at the front of the hauler where a steersman, perched on a small seat, performed the often-hazardous task of keeping the hauler and a long string of heavily-laden sleighs in the iced ruts, resisting the temptation to jump off and run on steep downhill grades.

Lombard's machines caused a sensation among northern loggers. They worked in weather that stopped even horses, and pulled a dozen or more sleighs, each with loads larger than horses could manage, over much longer distances at speeds of up to ten kilometers an hour. In 1906, a Canadian manufacturer, Alexander Dunbar and Sons, of Woodstock, New

“

*The banks of the Fraser River were not particularly adaptable to handlogging owing to the flat delta lands which bordered the waterways. It was soon perceived that the big source of timber would be Burrard Inlet which had the properly sloping banks to the water.*

*The felling, at first, was done entirely with axes, even for the large six, seven or eight–foot trees. The use of saws for felling came in the late '70s. Great skill was acquired by the axeman. The felling axe was developed, a double–bitted axe, with a long narrow blade, sharpened to a razor edge with a whetstone. Every axeman carried his own stone, which was a natural stone to be used with saliva to touch up the edge every few minutes. The edge was carefully ground, very slightly rounded so that the axe would not bind in the cut, but would come away freely. Proper use of the axe was somewhat like that of the use of a golf club or tennis racket where the whip given to the stroke gives it its greatest force at the point of impact.*

*The trees were felled almost invariable from the stump six or eight feet up, or higher, for two reasons: one was to get away from the gnarled or twisted grain near the ground in order to get a chip and split readily; the other reason, of course, was that from the greater elevation the better direction and the better jump or throw of the tree when it left the stump could be utilized. If the tree made a clean run to the salt water and floated, then all that was necessary would be to trim off the branches and buck off the top, and buck the tree into log lengths.*

*If, however, the tree lodged, then there was the problem of freeing it and getting it to run again. The tree would probably have to be barked. This was simple enough in the spring or early summer when the sap was running and the bark could be removed with a slice bar, which was something in the nature of a giant metal chisel, about one–half the length of a crow bar. If, however, the bark was tight then it was entirely an axe job. Great skill was required here. A log barked in this manner would give the same regular appearance as the scales on a fish. Then the front end of the log would have to be sniped or rounded so that it would ride toboggan–like over small obstructions. Then the log had to be started, and that was where the artificial power was used. This was almost solely the jack screw, by means of which the log or front end of the log could be raised to the proper height and started. Then it had to be blocked up and started with the application of force elsewhere. This was very often a dangerous procedure, particularly if a man was working under a log at the bow end, and when the bow was raised sufficiently to clear the obstructions it would often start to run and the logger would have to have a clear way to get out of its path.*

*Very often a log would plunge head first into the gravel or mud and lodge with the upper end still resting against the bank or bluff, as the case may be. For many years afterwards it was a common sight around Burrard Inlet to see these logs still lodged and leaning up against the cliff or bank. It was also common practice to try to free those logs. If they could not be jolted loose in some way, or by the use of the lifting effect of other logs attached at low tide to use the power of their buoyancy in lifting the log free, the only recourse they had was to cut it off. This, however, was a tricky and hazardous job.*

*My father tells of one time when he was offered a job by a chap he always described as "the white Frenchman," a handlogger of some renown. When they were about to start out to work the first morning, they were going to try to free some logs which were stuck in the mud, my father was asked if he had a rope for his axe. He was told that he needed a rope for the axe because when the log was cut off and started to run he would probably fall into the water and they could not afford to lose the axe. Needless to say, my father did not take the job.*

E.K. (Ned) DeBeck

Brunswick, also began producing a steam hauler, several of which were used in eastern Canadian forests. The steam haulers dominated much of the north woods during the first two decades of the century, with machines sold as far west as Saskatchewan and Montana. The Montana machines were used by the Western Lumber Company at Lathrop, where they were observed by the Holt brothers of California, who later adopted the endless track mechanism on their first steam tractors.

A few of these machines even made it abroad, including one to Russia. Word of their capabilities gradually spread through the northern forests and in 1912 Hugo Richart Sandberg, a forestry manager for the Kemi Company in Finland, heard that steam-engines equipped with tracks were pulling timber sleighs in Minnesota and Wisconsin. He thought they might be what was needed to haul logs over a ten-kilometer stretch to the nearest river in the Koilliskaira region of Finnish Lapland. During the winter of 1910–1911, he sent his engineer, Julius Miettinen to America to examine the machines and the methods. Upon his return, Miettinen reported that the steam haulers effortlessly pulled a long line of loaded timber sleighs. Sandberg ordered two haulers with spare parts and sleighs from the Phoenix Manufacturing Company in Eau Claire, Wisconsin. Phoenix had an agreement with Lombard enabling it to manufacture a modified version of the Lombard hauler, known as the Centipede, widely regarded as a superior machine. An elaborate facility was constructed at the Kemi operations in anticipation of the Centipede's arrival, including a full-service repair shop. Three pumping stations with water towers and steam pumps were built along the hauling route. The road bases were cleared and huge piles of fuel wood for the engines were stacked along the side of the road. About a thousand men and 250 horses were engaged to skid logs to the haul road.

The first engine arrived by boat at Kemi in the beginning of winter; from there it was transported by rail to the logging operations near Rovaniemi. There, the engine was dismantled and placed on special sleighs. The boiler alone weighed 7,000 kilograms and required twelve horses to pull it. At its destination, the engine was assembled and a dozen or so pairs of sleighs, each loaded with sixty to eighty logs, were hooked on behind it. The first load arrived at the river on January 16, 1913. Finland's initial experience with mechanized logging was underway.

Then, the difficulties began. The roads were built on terrain that was too hilly. The engines pulled uphill effortlessly, but going downhill, the sleighs piled up behind the engine, threatening to overturn it. Sandberg hired men to gather branches and reeds, and throw them under the runners to slow down the sleighs. So many brush men were needed that their own lodging was built. To prepare the roads, a herd of several hundred reindeer was used to pack the snow. Snow plows were mounted on the front of the engines and a water tank positioned on the last sleigh to water the road. Behind that a scraper was attached to cut grooves for the runners.

Although the Centipede boilers were designed for coal, there was none available in Lapland, so pine was used. Piles of wood marked the entire route, on both sides of the road. In spring, the sun shone on the soot from the engines and melted the road prematurely so the drives had to be stopped, even though there were meter-high snowdrifts all around.

It had been hoped that the engines would make two trips per day, but they barely managed one. During the first winter of steam hauling, only 23,500 logs were moved to the river. The following winter showed some improvement, with 45,000 logs transported, but by then the engines began breaking down. The engineer, Julius Miettinen, was not able to make all the spare parts needed in his backwoods shop. Getting parts from America was difficult, then, after World War One started, impossible. The costs of hauling were higher than anticipated, and by 1916, when it was shut down, the operation was showing a loss. But the seed of mechanization had been sown in Finland.

Oxen logging on East Thurlow Island, British Columbia, Canada, in 1890. Logs were skidded to this landing and rolled into the water with peaveys or jacks.

At the beginning of the century, logs were transported to the banks of rivers and lakes throughout the winter. During summer they were floated to sawmills located along the water routes, as shown on the Alimmainen Rautjärvi lake, Finland, in 1905.

Clearing jammed logs in 1910 in Scandinavia. If the jam could not be cleared with manpower, dynamite was used.

Pre-war logging camp in Lapland. Reindeer or men walking side by side were used to pack the snow on winter roads before water was sprayed on them from these sleds.

Lumberjack legends and songs provided entertainment in winter Scandinavian logging camps. Men toiled in the forests during the light of day and evenings were spent repairing equipment, eating and resting.

This form of hand logging, known as ball-hooting, was employed at Robbinsville, North Carolina, US, in 1956. Logs were rolled down steep ravines to roads below and hand loaded onto trucks.

Sarawack loggers in the late 1960s roll logs onto a sled that will be pulled by hand along the wooden skidway. This method was called kuda-kuda, or horse-horse.

Horses skid logs with a big wheel at this 1912 Michigan, US, logging site. The hardwood logs are then loaded onto rail cars with a McGiffert-style steam loader.

In Scandinavia during the 1920s horse-drawn sleds hauled logs to railways. This method was profitable for distances of up to seven kilometers, or slightly longer over frozen lakes.

Elephant logging in Myanmar in the 1970s. The log end has been sniped for easier skidding and a hole drilled, through which a skidding chain is inserted.

Lagerlöf's motorized sled, developed in Finland in 1911, was powered by a gasoline engine which generated electricity for a winch. The motorized sled pulled itself and a string of fifteen to twenty sleds along a cable.

A tractor logging site in Iisalmi, Finland, in 1925 utilizing a 35-horsepower Dinos crawler tractor.

A US-built Phoenix steam hauler in Northern Finland, in 1913. The 100-horsepower machine moved at 8 kilometers per hour, weighed 16 tons and consumed 12 cords of wood per day.

An early-model, steam-powered Lombard Log Hauler at work in the northeastern US about 1905. This was one of the first tracked machines built.

A Shay wood-burning locomotive hauling a log train in coastal British Columbia, Canada, in 1916. The Shay, built in Ohio, US, was the first locomotive built specifically for logging.

In 1912, this 26-ton Heisler logging locomotive was used in Arkansas, US, to haul railcars loaded with the steam loader in the background.

These Cletrac crawler tractors were imported to Scandinavia from the US in the 1920s. They replaced horses to pull log sleds on winter roads.

The Suojärvi truck logging site in Finland, 1927. Trucks replaced Dinos crawler tractors, which were not strong enough for the job.

A Dolbeer steam donkey used in British Columbia, Canada, in 1887. The spool-tender (right) maintains pressure on the cable and the vertical capstan pulls in a log.

A diesel-powered Lombard hauler appeared in the late 1920s and could pull 250 tons of logs on sleds over iced roads. For summer work, wheels replaced the steering runners on the front.

A steam-powered Clyde skidder operating at Oakdale, Louisiana, US, about 1910. Called a double-ended snaker, this machine skidded logs with each boom but did not load. It was self-propelled by chain drive.

A McGiffert loader lifts pine logs onto rail cars in the western US in the 1950s. Empty cars pass under the loader.

A McLean boom is employed to load a large Douglas fir log onto an early-model logging truck which will haul its load over a plank road in coastal British Columbia, Canada, in the 1920s.

Skidding logs with a steam-powered roader donkey at Fort Bragg, California, in the late 19th century.

# The Birth of Mechanization

By the 1920s, steam machinery had reached a pinnacle of sophistication — and the practical limits of its development for use in harvesting timber. In North America, several factors militated against further development of this technology. In some parts of the country, such as the West Coast, the easiest logging chances had been worked out and the available timber was now found on steeper, rougher ground, beyond the effective reach of steam railways. By late in the decade, it was apparent the great southern US pine forests were all but gone, and the regeneration of a second, new forest was scarcely imagined.

In the north, however, there was one area of potential expansion: into the spruce forests of northern Ontario and Quebec and, to a lesser extent, the northeastern US states. These forests contained timber ideally suited for pulp and paper production, but the trees were small and densely spaced. The available mechanical equipment was poorly suited for this kind of timber, so the only recourse was to revert to harvesting systems based on human and animal labor.

In 1918, to address this problem, the Canadian Pulp and Paper Association established a Woodland's Section, and ten years later hired A.M. Koroleff, an exiled Siberian logging engineer, as its first full-time staff member. A brilliant, energetic man — during his career he published more than one hundred books and articles on logging machines — Koroleff spent the next fifty years setting the pace around the world for research and development of logging equipment. The mandate of his group was to assist in development of pulpwood harvesting technology. Some years later, in 1934, with the rapid expansion of the pulp and paper industry in the US, the American Pulpwood Association was set up to represent paper manufacturers and wood suppliers. By the late 1940s, this group was also developing logging equipment. Together, these two organizations served as informational clearinghouses for the North American logging industry, apart from the West Coast, right up to the present. Similar organizations began to appear in Europe. During World War One, numerous studies on forestry were conducted in Sweden where they initiated active discussion and created a foundation for cooperative research between forest companies, forestry schools, universities, government organizations and the trade unions. In 1938, the major forest companies in western Sweden founded the Värmlands Skogsarbetsstudier (VSA) forest research institute with Gösta Luthman, an engineer, as its head and Per O. Olsson-Lokind, a forester, in charge of field work. The VSA quickly became Sweden's foremost forest research institute and, after World War Two, began work on logging equipment design and development.

The steady expansion of the global pulp and paper industry which began in the 1920s created a need for new logging equipment, a need that did not abate for decades, not even during the depths of the economic depression of the 1930s. A measure of the growth which occurred was reflected in the drop in Quebec pulpwood prices from $30 a cord in 1920 to $7 a cord in 1936, even though pulpwood consumption in that province doubled during the same period.

The mechanical innovation offering the most promise for a new direction in logging was the internal combustion engine. Internal combustion engines had a much higher power to weight and size ratio than steam engines, promising increased mobility in the difficult and variable terrains encountered by loggers. Throughout the decades of the 1920s and 1930s, a great deal of effort was expended adapting this new form of mechanical energy to the various phases of timber harvesting.

For loggers, the earliest application of the new engines was the electric generator. Previously, the rhythm of a northern logger's life was determined by the rising and setting of the sun. Through the winter months this meant short workdays and long, dark evenings. Portable electric generators dramatically changed this pattern, lighting up the bunkhouses, repair shops and outdoor areas. The workday lengthened, and shift work was introduced.

A few attempts were made to utilize portable generators in logging systems. In 1911, an employee of Finland's Raahe Timber Products Company named Lagerlöf built a sleigh on which a gas engine ran a generator used to power a winch. The sleigh pulled itself through the forest with the winch and, once in position, hauled loads of logs to the banks of a river, down which they were driven in the spring. Lagerlöf's experiments were interrupted by the Finnish civil war in 1917. At the beginning of World War One, a shipment of gas engines ordered for the new sleighs was confiscated in England. The trials were resumed in the winter of 1920–21 at Rovaniemi's Perunkajärvi site, where the new motor sleigh was used to transport more than 10,000 logs a distance of nine kilometers.

An important focus of North American mechanical development was the replacement of steam engines on cable yarding machines with gasoline or diesel motors. There were many drawbacks to steam-powered donkeys, including the large amounts of fuel and water they consumed. Initially, they burned wood, but the high wages required to employ wood buckers priced it out of the market. About 1917, due to manpower shortages caused by World War One, there was a widespread shift to oil, but the water problem remained. Attempts were made on the West Coast to convert existing equipment by substituting a gas or diesel engine for the steam plant, but the old winch systems were not fully compatible with the new engines and something better was needed. In 1921, Skagit Steel and Iron Works in Sedro Woolley, Washington, came out with the first gas donkey used in the northwest, called a Little Tugger. It consisted of a Fordson gas-powered farm tractor, the back wheels of which were replaced with sprockets that powered a set of winches. This machine, immensely popular among coastal loggers, was used extensively for yarding and loading, firmly establishing the Skagit company for decades to come.

In 1922, Skagit produced its first purpose-built gas-powered donkey, followed two years later by a diesel-powered model. Initially, these machines were used primarily as cold deck yarders, because they were lighter, and therefore easier to move to the back end of a setting, and did not require a supply of water. Several established yarder manufacturers, including Clyde and the G.B. Murdie Company in Victoria, British Columbia, switched production from steam to the newer type machines. Those that did not switch, such as Lidgerwood, soon lost their position in the logging industry.

For the next 20 years, until World War Two intervened, there was a rapid development — or redevelopment — of cable yarders using internal combustion engines. This was more than a reconfiguration of motive power; it also entailed a

> *They started using wire rope about 1900; it may be a little bit before that, but right around there. Before then, on the little donkeys they used Manila rope. I can remember when they first started using the wire rope. They used some wire rope and Manila for a long time, and then finally done away with the Manila rope, but it took a long time especially for the spool tenders to get used to handling this wire rope. You know, it was harder to handle, but it wouldn't break. Well, I don't know but that the other was probably as strong, but it was big and cumbersome, you know the Manila rope — two inches in diameter. And you would get as much strength out of a much smaller, a five-eighths wire rope. They had to do a lot of research work on that wire rope in order to get it to work right — took some time. And then they made it, built it a little differently. Finally they got it to work. No one liked it. Didn't want anything to do with it at all.*
>
> Frank Fraser

fundamental shift in thinking about the business of logging. Previously, the trend had been toward bigger, faster, more powerful yarders, the ultimate examples being machines such as the Lidgerwood Portable Logger or the Washington Flyer. Now, with motor trucks replacing railways, and the movement into higher, rougher ground and smaller timber, lightweight equipment had many advantages. The design emphasis shifted away from size to the development of more sophisticated transmissions, torque converters and other devices to make use of the new engines. A new kind of inventive genius was needed.

One of these was Frank Lawrence, one of the West Coast logging industry's most innovative and prolific mechanical virtuosos. Lawrence started designing logging equipment in the 1920s and a decade later came out with his first yarder, the gasoline-powered Lawrence 36. It was known, none too affectionately, as the Nutcracker because of the hazardous positioning of the strawline drum between the operator's legs. In 1942, he introduced the Lawrence 10–10, a three-drum, high-lead donkey with 350 meters of one-inch mainline powered by a six-cylinder Chrysler gas engine. For almost twenty years it was one of the most popular cable yarders on the North American West Coast. At the same time, Lawrence designed and built numerous pieces of logging equipment and components for others. One of his most enduring inventions was the Gearmatic winch, which eventually played a significant role in the evolution of an entirely new type of logging machine.

The internal combustion machine that most dramatically changed the nature of logging across the rest of the continent was the crawler tractor. This tractor, universally known as the Cat or Caterpillar no matter what its brand, was an outgrowth of two branches of tractor evolution — the steam hauler and the wheeled steam tractor used primarily in farming. The center of crawler development was in California, where a number of competing companies, including the Holt brothers and Daniel Best, an early builder of gas engines, set up plants. Both companies produced wheeled steam tractors used by western dry-land pine loggers. In 1904, the Holts produced their first track-laying machine, perhaps inspired by the Lombard hauler they had seen in Montana. It was a crude, forty-ton monster that steered with a front-mounted pilot wheel. It had one redeeming feature — it could outpull any wheeled machine on the market. Within a few years the Holts began using gas engines in their crawlers, and by 1910 about a hundred had been sold. Meanwhile, the Best company released a production crawler, the Model 40, a gas-powered machine with a four-cylinder, square-bored engine. In 1919, Best came out with its classic, Model 60, which dispensed with the pilot wheel and steered with clutches. A special model built for loggers was equipped with a fancy, if impractical, canvas canopy. From the outset, the Holt and Best companies engaged in a fierce competition and, eventually, lawsuits. Several years of litigation ended with the merger of the two companies in 1925, creating the Caterpillar Tractor Company.

“

*A high lead skidder, they called them prehaul skidders mostly in the South. Before my time over here, my understanding in South Carolina and North Carolina and all the swamps is that the high lead skidder started in the South and then migrated to the Northwest where they adapted what they were doing. A lot of the loggers in South Carolina and North Carolina during the start of logging out on the West Coast, they moved and went to the West Coast and logged for a while, then most of them came back. The first skidders that the old Allison Company in Belamy, Alabama, and my father used were steam skidders. Most of them were 2-drum or 3-drum, and you used trees for your pole. When I first came, they were logging with a 60-foot cyprus pole in a mud swamp. They had a little railroad went out in the mud swamp and stood up a 60-foot spar that they carried with them, and they'd pull it up. Took about a day or a day and a half to move the darn thing. They had a little spur line off the main line of the railroad and they got on this line and pulled the logs in. And then they loaded logs out with another skidder that was on a rail car.*

*In some of the areas they used piling. They built the railroad tracks ahead of them and with a skidder they drove pilings and built the railroad and went out through the swamp. In North Carolina, the muck was so deep that you couldn't drive pilings. They got the boys from South Carolina to come up there. What they had to do is just throw timber, cross up timber, and built a railroad on it and when the stuff sunk, you had to keep people out there all the time jacking it up and putting some more smaller trees under it to keep your tracks stabilized. And it sunk down, a lot of time a whole rail car would turn over out there and it was a mess.*

Joe Watkins

“

*Log skidding has been almost totally halted for four days now due to snow flurries and blizzard conditions. The latest snow flurries are the first natural obstacle to the skidding season.*

*The drivers have been able to turn the stoppage to their advantage, however. The horses lacking fodder will get some rest, and drivers can fix their defective gear. Some will have time to visit the mill and take care of other household chores. In this respect, the halting of log skidding perhaps feels welcome. There is the other side of the issue, however. In Ruovesi, for example, about 5,000 cubic meters remain unskidded each day of the interruption, so a four-day stoppage means 20,000 cubic meters. By the end of January, 112,667 m$^3$ of the targeted 290,000 m$^3$ had been skidded in Ruovesi. As soon as the flurries and blizzards end, the blocked roads can start to be opened, and skidding is expected to get back up to full-speed in about 24 hours.*

SOURCE: FINNISH NEWSPAPER *AAMULEHTI*, FEB. 8, 1946.

Through the 1920s a couple of dozen other machinery companies entered the crawler tractor business, most of them falling by the wayside during the depression and subsequent war. The major survivors were Allis-Chalmers and International Harvester, neither of which played a particularly significant role in the logging industry.

The uses to which loggers put their crawlers were as varied as the conditions encountered. Initially, crawlers replaced other sources of skidding power, particularly the horse. In areas where ground skidding was common, logs were simply dragged along behind the tractor. Elsewhere, where big wheels and other devices for raising logs off the ground had come into use, tractors pulled these devices. In the West and the South, an elaborate tractor accessory technology quickly developed. Pans were added, then bummers. With the addition of winches to the rear of the tractors, the big wheel evolved into a wheeled arch, under which the logs were raised with the winch line. The Athey Truss Wheel Company of Chicago produced a tracked bogie which became immensely popular on arches, providing the ability to skid over soft ground. Double-drum winches were used in several situations — as high lead yarding engines, to raise spar trees, as loading engines, to operate draglines — limited only by the imagination of their operators.

The adaptability of the new crawlers was enormous. With the addition of fairleads to the top of arches, winches were used to yard logs for thirty meters or more, before they were bunched and skidded with the tractor. They climbed steep slopes and skidded logs down slopes of sixty percent or more. Their power was sufficient to handle more than a few logs in a turn, which led to the skidding of longer and, eventually, tree-length logs.

Crawlers were also used in conjunction with, and to extend the life of, other logging systems. In many parts of the West Coast and in the western and Appalachian mountains, they replaced cable yarders. Yarding downhill with powerful steam yarders was resulting in heavy log breakage. As a result, cable yarders, and the railways to which they skidded logs, were relocated to the high ground along ridges and hill tops — a difficult and expensive proposition. The use of crawler tractors meant that logs could be skidded safely down steep slopes, allowing the return of the railways to the valley bottoms where they operated more easily and economically.

Elsewhere, crawler tractors simply replaced other methods. One of the earliest victims was the steam hauler used to haul sleigh trains in the northern forests. Similarly, horse-drawn wagons became tractor drawn. Tractors replaced horses and roaders in skidding logs down chutes. In the late 1930s, eastern Canadian pulpwood companies asked the equipment manufacturers to find alternatives to shortwood cut-and-pile pulp logging. One of the first responses to this appeal was the logging sulky, a rubber-tired arch well suited to tree-length logging introduced in 1938 through a collaborative effort between Caterpillar and Hyster. This type of development, along with other adaptations of the crawler tractor, revolutionized logging throughout North America and in many other parts of the world.

One of the most dramatic additions to the crawler tractor's capability came from a simple accessory. It was invented by a California highway contractor who had built his own earth scrapers and was firmly convinced that his partner in these enterprises was God Himself. Bob LeTourneau was one of the most eccentric, prolific and creative industrial machine designers and manufacturers the world has ever seen. His solution to the problem of filling a deep wash, on a contract he held building the Oakland-Stockton highway in 1926, was to hang a dozer blade off the front of a Best crawler tractor. People had tried putting blades on tractors before, but they had met with little success. LeTourneau's invention included a means of raising and lowering the blade with a line from an electric cable winch. A touch of a switch controlled the height of the blade. This invention gave the crawler tractor the ability to build roads.

A dozen years or so before LeTourneau designed his bulldozer, motor trucks made their first appearance in the woods, probably in Washington or Oregon. The trucks of that period were primitive affairs, with small engines, steel or hard rubber tires and flimsy frames. They were intended primarily for use on well-graded or paved roads. In the woods there was only one practical way to provide the kind of running surface a truck needed — build it with timber. What quickly evolved was the fore-and-aft road, a structure consisting of an underframe supporting flat hewn logs placed end to end and spaced apart the same distance as the truck wheels. Early attempts with this equipment involved loading logs directly onto the back of the truck, but only short logs could be hauled. Soon, trailers were devised for the truck to tow.

Priced halfway between a horse and a locomotive, trucks quickly moved into the mid-ground of log transport, taking logs from the skidder or horse landing to the railway or, even, the dump or mill. A real boost to truck logging occurred

"

*I was a bullbucker at International Timber and on New Year's day 1928 Pete Harambourne came up to Camp 7 in quite a sweat to take an inventory of cedar poles piled away up the track. It was quite amusing. He looked for everyone in the office and scaler's shack for someone to do this work, passing me by with scarcely a glance. Pete just could not find anyone that was able to do this work. It was a Sunday anyway, and most of the guys with a touch of the morning-after were not interested. So there was nothing to do but ask me. I said okay, how do I get up there. He said, "I'll send you up on the Baldwin mainline locomotive, and he will stay with you until you are finished." So, away I went.*

*The work was some distance from camp on a newly laid spur road. There was just myself and Bill Johnson, the engineer. Going over this newly laid spur road that had practically no traffic on it, the big locie, for some reason, spread the tracks and settled right down in the mud, all wheels off the rails. We were not far from where I had to be, and Bill said, "You go ahead and I will try to get her on the tracks." So I went to work out of sight of the Baldwin. I guess I was away about two to three hours, and when I came back Old Bill was sitting in the cab with the fire door wide open, and the locie still mired down between the tracks.*

*Bill said, "It's no use Arch. This old girl is settled down for the night, and we might as well stay with her. It will take a section crew half a day to put her on the tracks, and when we don't show up Pete will be up here raising hell." Well, Bill was a guy with quite a sense of humor and not given to getting excited over anything. So there we sat, with the windows shuttered off with the side flaps and the fire door open. Since it was New Years, Bill had fortified himself with a mickey of rum. I heartily commended him on his foresight. We stayed there till past suppertime and, sure enough, Pete himself came looking for us. After his usual blast when things went haywire, Bill poked his head from behind the curtain and said, "Pete. This is one hell of a railroad you have — it is not the longest in B.C., but it sure is the widest."*

Archibald Kerr

> *They started developing cable winches for the rear of the 'dozer for logging, and the arch came into use in the middle '20s, I imagine, or late '20s. It gave them the opportunity to reach out with the lines and pull the logs to the tractor road rather than getting the tractor stuck off at the side of the road. The other benefit it gave the tractor was that if it did get stuck, it didn't need another tractor to help it get out again. It could winch itself out of a place where it got mired down in the soft places.*
>
> HOWARD PETERSON

near the end of World War One when the US government purchased 150 Standard trucks to speed up production of airplane spruce on Washington's Olympic Peninsula. Elsewhere, trucks were quickly adopted wherever adequate public road systems were in place — notably the New England and Great Lake states, and in Sweden. In Michigan and parts of the South, plank roads were commonly built into logging operations. And in the North, trucks were used on ice roads, pulling log-loaded sleighs. In one instance, during the winter of 1928-29, the Ahonen Lumber Company of Ironwood, Michigan, pulled loaded sleighs with a powerful Pierce-Arrow passenger car. The rig worked well and operated at great speed along a state highway until it was banished by the police.

During the 1920s, there were two to three dozen makes of trucks available, all rated at under two tons and most with a single set of rear wheels. Loggers habitually hauled loads well in excess of the truck's carrying capacity, not to mention its braking capability. There were some early attempts to produce heavy-duty trucks for logging and other heavy work. Lombard, the steam hauler company, and Linn came out with a cross-breed of a crawler tractor and a truck, intended for use on ice roads. Each had a hundred horsepower gas engine, tracked drive wheels, bobsled steering and large flat decks fitted out with log bunks. They were not a success.

Being adaptable beyond all else, loggers tried trucks in every imaginable configuration. They were fitted out with special wheels for use on log rails, and equipped with railway wheels for use on standard and narrow gauge railways. In the West, with its big logs and rough ground, development of prototype heavy-duty logging trucks began with the establishment in 1922 of the Hayes-Anderson Motor Company in Vancouver, British Columbia. It lasted more than 50 years, specializing in the manufacture of logging trucks, considered by many loggers the finest ever made. In its early years Hayes-Anderson racked up an impressive list of "firsts" in logging truck design: first dual axles, first use of diesel engines, first self-loading trailers. For a few years, while Frank Lawrence was a partner, the company was called Hayes-Lawrence.

The year after Hayes set up shop, John Holstrom started the Kenworth truck company in Seattle, specializing in custom-made logging trucks from 1928 on. Apart from these two companies, most truck builders were located in eastern North America. The four Mack brothers set up a company in Brooklyn in 1903, and a dozen years later came out with the Bulldog, a popular logging truck built until 1939. The White Motor Company of Cleveland, Ohio, started out building sewing machines, branched off into steam-powered automobiles, developed the Cletrac crawler tractor widely used by loggers, and in 1913 began making what eventually became, arguably, the most popular logging truck of all time, the White.

The greatest limitation on the expanded use of trucks stemmed from the related problem of tires and roads. LeTourneau's bulldozer contributed enormously toward the solution of the latter problem by making road building a much easier, cheaper task. A logger could buy a crawler tractor and equip it with a two-drum winch, one to skid logs, the other to operate a dozer blade. Only one, dual-purpose machine was required. On the West Coast this combination was used effectively in conjunction with one of the new gas or diesel cable yarders, which were compact enough to be moved over the dozer-built road by a logging truck.

Tires were another matter. Until the hard rubber tire era ended, the only practical solution was the plank or fore-and-aft road. Pneumatic tires appeared in the early 1920s, but they lacked the strength needed for logging until Goodyear came out with a 12-ply, rayon cord tire in the early 1930s. This improvement coincided conveniently with the widespread availability of bulldozers, which could build the type of roads pneumatic tires could handle.

The adoption of trucks created a major problem with log loading. In some cases it was solved by reverting to more primitive methods, such as hand loading of pulpwood, which could be cut to short lengths to make it manageable. If horses or tractors were used for skidding to roadside, trucks were loaded by cross hauling. This method evolved into the development of "jammers". These were essentially A-frames, some with booms, which were powered by horses or tractors and lowered tonged or hooked logs onto the truck bed. In the West, cable loading systems used to load rail cars were also used to load trucks, often with disastrous consequences for the truck. A rail car stood up to most of the rough treatment that loading West Coast logs entailed, but most trucks didn't. A lot of loader operators, who had learned their trade on a steam machine loading rail cars, found themselves headed down the road after dropping a one-meter-diameter Douglas fir log on the back end of a new White or Mack truck. The solution to their difficulties had to wait until after the next war.

In the pre-war years there was one last bastion of non-mechanized work in the woods which had eluded the machine makers — the falling of trees and processing them into logs. During the 1930s, a labor shortage developed in most timber-producing areas of North America, outside the southern pine region where a large rural workforce was still available. Untrained European immigrants had provided most of the continent's heavy manual workers, including six out of every seven of those employed at the strenuous task of falling and bucking timber. In the late 1920s, this supply of immigrant labor dried up, just when the need for fallers and buckers was increasing. The northern pulp region was expanding its operations, and on the Pacific coast the available timber was becoming smaller, reducing individual production rates.

Many larger forest companies began to look at some of the mechanically operated saws then under development. In 1930, the CPPA Woodland Section undertook a survey of the power saws available. At that time about a dozen motorized chain saws were on the market. The concepts of both a power saw and an endless cutting chain were not new. A form of saw chain was invented in Wales in the seventeenth century for use with hand-cranked machines to cut pit props in coal mines. It had been used for that purpose in the US since the 1880s. During the last half of the nineteenth century, twenty five US patents for endless cutting chains were issued.

The development of suitable engines was a slower process. A patent was issued to a New York designer for a chain saw in 1858, and several more to other inventors over the next few decades. The invention of the electric engine in the 1890s led to the development of several electrically

"

*The unreliability of the motors in the early days was a big problem. We had lots of down time. Due to vibration the ignition wires would break off. There were no air filters on the breathers. Sometimes, if the operator wasn't careful, he'd get it in the pile of sawdust and it would suck in the sawdust and seize the motor up. You'd take your tools and take the cylinder head off, chip out the sawdust in the cylinder and go to work. Today they're so highly machined a bit of dust will stop them. In those days they were pretty crude and they could be monkey wrenched on the job quite easily.*

*The early fallers weren't mechanically sophisticated at all, to put it mildly. Myself, raised on a farm, I knew a little bit about machinery. For most of them, to be put on a machine, it was baffling to them. In those days we had some spares with us. But if you broke a chain you'd have to fix it in the field yourself. Repairing a broken chain on a stump, when the no-see-ums and mosquitoes were killing you, took a long time. Now, we'd never ask anyone to repair a chain on the job. Repair it in the shop. Now the cost of chain, in the whole cost of keeping a falling and bucking crew going, is nothing.*

*We did most of the repair work in the shops. Of course, if you were working on contract and the saw broke down you didn't get paid. You had to try to improvise. In the start there were still axes out in the woods, and maybe a handsaw as a spare. You could finish off the day with a handsaw.*

Olaf Fedje

“

*It was a very, very competitive sort of an industry (in the 1930s), individually and in groups. In the falling and bucking of timber, it was done on what they called the bushel basis. They got paid so much per unit, or so much per thousand; and naturally this stimulated a lot of incentive for these fellows. They would use all hand tools. You would see these fellows as soon as it was daylight start in with a saw, and they sometimes wouldn't even stop for lunch. You'd see them with a sandwich in one hand and a saw in the other. They would work this way right up to the time that transportation would come along to pick them up or they'd have time to walk out to catch the transportation.*

*In the rigging there was always this competition between hook tenders, who are the foremen of each logging site, to come in at night and brag about how many more logs he got than the other side did. Now along with this, practically every hook tender had his own crew. If you were working in a crew and a new hook tender came along, you might just as well figure that you'd better pack your bag and get out, because he was going to bring in his team, you might say, just like a coach would bring in all his players.*

*In those days there was always one in the crew who they called the bull choker setter, and he was the one who set the pace and they got a dollar more a day. If you went into a new crew, if you had any experience, you could soon tell who was the bull choker setter. And it was always your desire to beat him to the choker because you knew that if you beat him often enough during the day, you'd either get his job or you could go talk to whoever was responsible for running the camp and say you wanted a dollar more a day. If they said no, you usually quit.*

*But most of the time if you had proven yourself, you'd get the dollar more a day or four bits more a day, whatever it would be. And this wasn't the best situation because maybe this next fellow would get bumped down and may even lose his job over it. This was typical. And of course, it went right on to the point where they had to see who was the best man when it came to scrapping and fighting. This got exaggerated too, you know. And these fellows began to believe what was written about them. It didn't help the situation too much.*

HOWARD PETERSON

powered saws, and in 1905 the Ashland Iron Works in Oregon manufactured an air-powered chain saw. The Potlatch Lumber Company, in Idaho built a deck-mounted chain saw in 1906, which inspired a Spokane inventer, Charles Wolf. He designed and, in 1920, began building three models of a semi-portable electric saw at the Peninsula Iron Works in Portland. The Reed Prentice Company of Massachusetts acquired a license to make these saws. A California water pump manufacturer, Dow, sold a 225-kilogram, gas-powered chain saw mounted on two motorcycle wheels. By 1930 the Beaver, a falling saw with an endless chain held by a buck-saw type frame, was being sold in Connecticut.

Coincidental with these developments were a number of similar ones in Europe, based on superior engineering of small gas engines. Sector, a Swedish company, began building two-man saws in 1916. Radically different in design, they consisted of a motor connected by a meter-long driveshaft to a U-shaped saw head with rollers and an enclosed cutting chain. Another Swedish saw, the Gerber, was a more conventionally designed machine which came on the market in 1919. The Arbor, an overweight, underpowered Finnish saw built in Turku, was on the market from 1916 to 1920. One of the first European saws to find wide use in North America was the Rinco, built in Luban, now part of Poland. It had a high power-to-weight ratio, used a modern-type cutting bar and the engine could run in a horizontal position for falling, probably the first chain saw with this capability. An Austrian saw of similar design, the Sylva, was built by the Vitak and Rainer Company in Vienna.

But apart from the Rinco, the most advanced chain saws of this period were built in Germany. One of the most enduring of these was a saw developed by the inventor Emil Lerp in 1927. He called it the Dolmar after a mountain in the region where he tested it. Lerp and Alfonso Lang founded the Dolmar Machinfabrik at Hamburg in 1930 following the early

> *Most of the time what they did in the big cyprus swamps, they went in and girdled the cyprus. Took men with axes, stuck in a pole to stand on, then they'd cut this tree all the way around, cut the cambium layer so the tree would die. And they let those trees dry for about a year, or two years sometimes, then they would go in and cut them down in low water. At high water they'd float them out and make a raft. They just brought them out of the swamps on high water and wired them together and floated them into the sawmills down the river.*
>
> JOE WATKINS

success of this saw. In their first year they offered seven models of chain saws, four of which had electric motors. The bar on the Dolmar saws swivelled to permit both falling and bucking with one machine. Another German saw, perhaps the most successful chain saw ever built, was the Stihl, which first appeared in 1926 when Andreas Stihl put an electric chain saw on the market. In 1930, he came out with a gas-powered saw, which met with great success in several parts of the world. In 1932, Stihl received an order for three thousand saws from the Soviet Union, and was told he could sell five thousand in North America if he could deliver them.

In addition, a number of other mechanical saws not utilizing cutting chains had been devised and, in a few cases, marketed. Some of these devices were only accessories to assist in the operation of hand saws, such as a Swedish coil spring called a "kompisapparat" used to pull a crosscut saw back through the cut. Fallers on the American West Coast occasionally utilized the same principle by attaching one end of their saws to a bent sapling.

One German inventor designed an electrically heated platinum wire to burn through trees. A rasp mounted on the end of a rotating shaft, intended to chew its way through a tree, was tried. Another machine used an auger to bore a series of holes through the tree. Various reciprocating saws were built, powered by steam, air or internal combustion engines, and utilizing modified crosscut blades. Some of these drag saws, as they were called, were widely used on the West Coast in stationary arrangements to buck firewood for steam donkeys. But, for the most part, few of these devices were of any practical utility. And, in 1930, neither were any of the chain saws on the market — with the possible exceptions of the Dolmar and the Stihl — as far as most loggers throughout the world were concerned. They were too heavy, cumbersome, unreliable and expensive.

By 1936 the situation began to change, largely because of a labor shortage. That year the Swedish Bergslags Company converted its Stora Kopparbergs logging operation in northern Uppland to chain saws. Coincidentally, the British Columbia Loggers' Association and the US-based Pacific Logging Congress launched a joint effort to develop a saw suited to West Coast conditions. After much experimentation, a number of Stihl saws were imported from Germany. Bloedel, Stewart & Welch, a Bellingham, Washington, logging company with extensive timber operations in British Columbia, undertook to test the saws and work with Stihl to modify them. The job was turned over to company bullbucker, Jack Challenger, who began training workers and making improvements to the saws at the company's Franklin River camp. A fundamental problem in introducing the new saws was that most experienced fallers lacked the mechanical experience required to operate and repair them. As well, experienced fallers were disinclined to lug the 55-kilogram monsters around in the woods. The solution lay in hiring husky young farm workers from the depression-wracked Canadian prairies, who had experience with gas engines. Called "machine men," they carried the heavy, motor-end of the saw, operated it and kept it running, while the fallers managed the relatively light end of the 1.5-meter bar and controlled the actual cutting of the trees.

By the fall of 1939, Bloedel, Stewart & Welch's power saw fallers' annual production was up to 500,000 cubic meters.

Just after the company placed a large order for more Stihl saws, war was declared with Germany and all trade with the country was cancelled. Immediately, D.H. Smith, the Stihl dealer in Vancouver, began building a replica of the Stihl, as did the Mall Tool Company in Chicago a bit later. During the war, Smith sold his operation to Reed Prentice, then set up the Hornet power saw company in Ontario, which he later sold to Mall. Reed Prentice soon sold Smith's original Vancouver company to a group of employees, who renamed it Industrial Engineering Limited, commonly known as IEL. Under this name, the company became the largest manufacturer of chain saws in North America, if not the world, for the duration of the war. Wartime labor shortages created a huge demand for the new saws, prompting the establishment of several manufacturing companies across the continent during and immediately after the war. One of these, the Poulan company, was set up in 1945 in Alto, Texas, by Claude Poulan, who also invented the bow guide or bar, which, after the war, took the southern pine region by storm.

With the exception of chain saw development, the advent of war brought the mechanical evolution of logging equipment to an abrupt halt. For at least five years, the industry was frozen in its 1930s mode, a mix of steam and internal combustion machinery. The war itself produced many technical advances which eventually found application in logging machinery, but until it ended there was no possibility of continuing the evolutionary processes which had begun during the previous two decades.

The technical developments that took place during the 1920s and 1930s — basically, the introduction of motor trucks, crawler tractors and chain saws — had an enormous impact on the industry. In many regions, particularly in the northern boreal forests of Europe and America, logging was transformed from a rural occupation that shared characteristics with farming and fishing, to an industrialized occupation closer to that of the factory. By the end of this period, logging had become deseasonalized. The transporting of logs was no longer dependent on heavy snowfalls, freezing temperatures and a spring freshet; trucks hauled year round and through the night. Logging became more of a full-time occupation and attracted a more dedicated type of worker, who often drove to work each day and returned to his family in the evening or on weekends. A workforce of professional loggers began to emerge.

These technical changes not only transformed the fundamental nature of working in the woods, but exerted an enormous influence on the industry's structure as well. Steam technology tended toward bigness, not only big, powerful machines, but big operations as well. Price of entry into the railway logging business was prohibitive to all but well-established operators and large corporations. It was difficult for a small horse logger to become a railway logger, and few managed it. Railway logging, as well as inflicting enormous destruction upon North American forests, was also unhealthy for the development and maintenance of the kind of free enterprise economy based on individual initiative and innovation much needed in an evolving industry such as logging. The introduction of these new technological devices, however, offered smaller logging operators vastly expanded opportunities. A truck, a small tractor and a chain saw were within the financial reach of an individual or a small firm. Eventually, when the war ended, the full impact of these new machines on the character and structure of the industry would be fully appreciated.

Horse logging was by far the most common method in Scandinavia during the 1930s. Even into the next decade, there were more than half a million men and two hundred thousand horses working in Scandinavian forests.

Handfallers use a crosscut saw to fell a Sitka spruce on the British Columbia coast prior to the development of the chain saw in the early 1940s. The fallers have chopped holes in the tree with axes to insert springboards upon which they stand to chop an undercut before felling the tree with a saw.

Bucking a pine log in the southern US in the 1940s. An axe has been driven into the log, and its handle used to support the saw.

Hand-saw tests underway at the Evo Forestry School in Finland in 1937.

Fallers at Vancouver Island, Canada, demonstrate the use of an early IEL chain saw in 1940. This saw was a wartime copy of an early Stihl two-man saw.

A demonstration to Finnish loggers in the late 1920s or early 1930s of what is probably a two-man Rinco chain saw.

A Pioneer bow saw is used for bucking pulpwood in the US South. This type of bar reduced pinching when bucking trees lying flat on the ground. The short logs are being hand loaded onto a truck which has been driven into the woods.

A Scandinavian cutter-type debarking machine in use in 1928. Pulp was not bleached at this time, so pulpwood had to be clean.

A debarking machine for pulpwood developed by Mauri R. Lagus, the builder of Finland's first chain saw; 1928.

A portable German debarker in 1936. More than half of a logger's time was spent debarking.

A four-wheel drive Duplex logging truck skidding logs along a plank road in coastal British Columbia, Canada, in 1924. The rear wheels were wrapped with cable for additional traction.

A Skagit Little Tugger ground skidding logs in the Pacific Northwest in the 1920s. The machine was built from a Fordson farm tractor, and mounted on a log sled. The rear wheels were replaced with sprockets which drove the cable winch.

A Hyster high-lead winch mounted on a Caterpillar D7 is used to skid logs at Sumner, Washington, US, in 1943. These rigs were used in several parts of North America after World War Two.

A Caterpillar tractor towing a steel-wheeled arch is used to skid logs in the United States in the 1940s.

Pioneer Timber Company loggers pose with their White logging truck in 1936, on a fore-and-aft road at Malcolm Island, British Columbia.

A truck fitted with railroad wheels was used to haul logs in the southern US during the 1940s. In this post-war period many ingenious adaptations were employed by loggers.

Hand loading trucks at the Finnish Kuolajärvi logging site in 1937. The truck was owned by Väinö Eskelinen.

Mechanized loading started appearing in Finland in the late 1930s. This machine, powered by a 35-horsepower Opel crude oil engine, could load a truck in twenty minutes.

After World War II in Finland various machines were adapted to log. Here, a Finnish shop-built logging arch mounted on runners is pulled by a US-made crawler tractor. Logs were attached to the arch with chains.

A Holt crawler towing an Athey bummer is used in the early 1920s at the Stillwater logging camp on the British Columbia coast to skid Douglas fir logs.

In Scandinavia, farm tractors equipped with loading booms were used for loading logs.

Logs were loaded with the help of tongs. In winter, this could be dangerous.

Even though machines pulled the logs, rolling a log onto a sled or to within reach of chains required skill and hard work.

In Scandinavia, crawler tractors were used as loaders and for stocking logs at roadside.

# Post-war Recovery

Globally, the forest industries that emerged from World War Two were in a much different condition than they had been a decade earlier. In parts of Europe and Asia, the forests themselves were severely damaged by the war, and were pillaged even further during early reconstruction phases. Like the rest of the economies of these areas, the forest industries were shattered. Elsewhere, though, the situation was quite different. In the Scandinavian countries, the Soviet Union, North America, Australia and New Zealand, unscathed forests were available for exploitation. There was also anticipation of an enormous increase in global timber demand.

North America experienced a housing boom to meet the needs of returning veterans, plus expanded opportunities to supply building materials for European reconstruction. Perhaps most pressing was the explosive demand for wood pulp. The war had created a multitude of new uses for cellulose fiber — fabrics, plastics, chemicals — as well as an expanded need for paper. In the US South a new, second-growth pine forest, which had grown up to replace the cut-over natural forest, was ideally suited for the harvesting of pulpwood. And there was capital available, accumulated during the war years, to finance the construction of numerous pulp mills throughout the South, as well as in eastern Canada.

The problem was the state of the logging industry. In North America, apart from the West Coast which had come out of the war possessing a mix of steam and internal combustion powered equipment, the rest of the industry was still based on the muscle power of humans and animals. In the South there had even been a mechanical regression after the last of the great pine forests had been logged out in the early 1930s. The steam-powered logging infrastructure had all but disappeared and had not been replaced with new logging systems built around trucks, tractors and gasoline donkeys.

Throughout the great pulpwood forest stretching from northern Ontario and Quebec east and south through the Canadian maritime provinces along the Atlantic seaboard to Florida and east to Texas, a similar system of logging prevailed. Men felled, limbed and bucked trees with hand saws and axes. If they were producing sawlogs, they skidded them to roadside or a landing with horses or mules. If they were cutting pulpwood, they bucked the logs into manageable lengths and loaded them by hand, often several times before they reached a railway or a driveable river. To make matters worse, it had become difficult to find men able and willing to work so hard, or who knew how to work with horses and mules. During the war, an entire generation of young workers had gained experience with the most advanced mechanical equipment. By the end of the war, farmers, who formed the core of the forest labor force, had sold their draught animals and bought tractors. Only the least skilled members of the labor force were willing to go back into the woods and put up with the low wages and difficult working conditions found there. The best workers wanted higher paying jobs which demanded more from them than mere physical exertion. Better jobs were available in the new factories and manufacturing plants in rapidly expanding urban areas. Whether they were Lake-state farm boys, French-Canadian woods workers, Appalachian mountain boys, southern black field hands or maritime fishermen, they all felt the same — they were not going back to the woods if they could avoid it.

The war itself did not contribute directly to the mechanization of logging. Apart from the wartime development of power saws, there were practically no advances in logging equipment during the war. There were many technical advances, of course, in fields such as metallurgy, diesel engineering, hydraulics and plastics, all of which eventually found application in logging equipment. But at war's end these improvements lay in the future. The logging industry emerged from the war in essentially the same state of mechanical development as it went into it.

Even during the last half of the 1940s and into the early 1950s, very little mechanization occurred in the woods. What did take place, with one notable exception, was localized adaptation of available machinery. In farming areas, such as southern Michigan, rubber-tired farm tractors with gas engines were pressed into service. They lacked the necessary power, had poor traction and maneuverability, and were apt to roll over on rough ground, but they were available. So was a wide range of trucks, all of which had been built for easier conditions, but upon which an overload of sawlogs or pulp wood could be piled. In the South, loggers made extensive use of surplus military trucks well into the 1950s. Crawler tractors were perhaps the most useful machines around. They were slow and their tracks were prone to breakage around stumps and rocks, but they were better than anything else available.

On the North American Pacific Coast, the situation was quite different. Cable logging was still the primary system in use, and after the war its conversion to internal combustion power resumed, as did the transition from railway to truck logging. In the West, machine manufacturers dedicated to the logging industry were still in business and able to build the machines they had dreamed up during the war years. By the end of the war, Kenworth and Hayes had been building logging trucks for twenty years, and Skagit had been manufacturing modern cable yarders for almost as long. Still, the established manufacturers looked upon the logging industry as a minor market. Truck makers such as General Motors or Ford, and tractor companies such as Caterpillar, were not sold on the idea of developing machinery dedicated to logging. This attitude was dramatically revealed just after the war when Tom Flynn of the US Forest Service designed the Tom Cat, a small log-skidding tractor. It was a tracked vehicle with a winch that hoisted logs up onto its sloping rear deck and skidded them to roadside. North American tractor manufacturers responded to requests to produce this machine with the uniform attitude that the general purpose tractors they were already building could be adapted to the task of skidding logs.

The one area in which mechanization did proceed was with the rapid evolution of the chain saw. Prior to and during the war, use of gas and electric powered chain saws mushroomed on the West Coast. The first of these saws — Stihls, which initially were imported from Germany, then manufactured at various locations throughout North America — were cumbersome, heavy two-man machines. In big timber it was worth the effort to lug one of these machines into the woods. In the dense spruce forests of eastern Canada, in the smaller-stemmed pine forests of the South and in the southern swamps, they were impractical and, consequently, not widely used.

During the war, many advances were made in small-engine design, and several manufacturers gained experience building and using new, lightweight, air-cooled engines. One American company, McCulloch, developed extensive wartime expertise building two-cycle engines to power aerial target drones, as well as engines for Reed

“

*When I came to Mobile in 1965 as manager of wood procurement they had one logging crew with fourteen people in it. It never produced fourteen thousand cords of wood a year. It never got up to the level of a thousand cords per man. It was a pallet operation. Shortwood. Cut down the wood, put it on a pallet. Skidded that thing out into the woods with a D3, then loaded it and pulled the pallets up on a truck. And I don't care if you had five thousand pallets in the operation, sooner or later they either all got full or they all got empty, or they all got in the wrong place, and the whole damn thing stopped while you went and got 'em. There probably were cavemen logging with dinosaurs who had more inefficiencies built into their logging operations than that one, but nobody since the glacier receded had a logging operation that inefficient.*

Tom Kelly

Prentice saws. When Reed Prentice left the business, and the war ended, McCulloch began producing its own saw, which quickly evolved into a light, 11-kilogram, one-man saw that swept the market in the 1950s. Other manufacturers — Titan, Homelite, Mall, IEL and a score of other North American companies — were producing lighter saws equipped with a radical new chain devised and sold by Joe Cox from his Oregon Saw Chain plant in Portland. These saws were designed and built specifically for loggers, to fall, limb and buck trees. By the mid-1950s, they had replaced almost every axe and handsaw in North America and were making serious inroads in Europe. The first lot of one-man American saws appeared in Swedish stores in 1946–47, and interest in them increased in 1948 when the Canadian Hornet and the Norwegian Jo-Bu were introduced to the market. That year, 600 chain saws were sold in Sweden. In 1950, the first Swedish one-man chain saw was developed, the Bebo, and the battle for market share was on. Competing in the race were the Swedish Tornado, Typhoon, Husqvarna, Raket, Partner, Comet and Companion. West Germany resumed exporting Stihl and Dolmar saws, and from Britain came the Danarm. The state-owned Valmet, in Finland, produced an electric saw, the Kita, for shipment to the Soviet Union as a war indemnity, but it was never a success in its home country.

It was one of the most revolutionary mechanization processes in the history of the industry. Workers who used the new saws — fallers, choppers, flatheads, pulpwood cutters — took to them eagerly. The reason was simple: most of these workers were paid piece rates, and the new saws substantially increased productivity and, therefore, incomes.

In the snow-belt forests, saws with the so-called beavertail bars were favored, but in the American South, where fallen trees had to be bucked while lying in the dirt, Claude Poulan's bow saw was more popular. It was easier to keep out of the dirt and was not so easily pinched in the cut. Also in the South, the chain saw had to compete for a decade or more with another type of power saw, the wheel saw. This machine first appeared around 1940. It consisted of an air-cooled gas engine mounted on two pneumatic tires, connected by drive belts to a half-meter circular saw blade which could be set in a horizontal position for falling a tree, or vertically for bucking. (Limbing still had to be done with an axe.) The wheel saw — produced by companies such as Lowther of Chicago, or Kut-Kwick and Tidewater Equipment, both of Brunswick, Georgia — had two handles. Operators pushed it around like a wheelbarrow over the relatively flat pine forests of the southern US coastal plain. It had very little utility outside the South, and was probably responsible for more injuries than any machine ever used in the woods. The large blade easily kicked the saw out of the operator's hands whereupon it swung around and cut off his leg at the ankle. But the largest number of injuries was caused by workers starting the engine with a tug on the drive belts and losing their fingers in the pulleys. Although it was still used, mostly for bucking, into the early 1960s, the wheel saw was eventually driven from the forest by the chain saw.

“

*We were always gyppo loggers. We bought three military surplus Mack trucks that were just borderline highway-size trucks. They had old-fashioned Buda diesel engines in them that were real dogs. We took out those and put in our 671 Jimmys and they'd run rings around any of the trucks that were here in that era. Why I have a hearing problem now is because of those things. They screamed, made a terrible noise. In those days people weren't concerned about ear difficulties. We just had a straight pipe up the back of the cab. They drummed in the cab, and just about deafened anybody at the side of the road when you went by.*

*Later we got hold of some of those monster Mack trucks that had been used on the St. Lawrence Seaway. They were real big ones. We bought two of those and converted them to logging trucks. They had a V-16 Cummins diesel engine in them. They packed a lot of wood. They were pretty heavy on maintenance. There were a lot of parts in the rear-end assembly that required a lot of maintenance. We eventually sidetracked those and got into Pacific trucks.*

Frank Norie

“

*About 1949 or '50 I was working for a lumber company in Montana where all the trees were felled with hand saws. Two-man saws, about six feet long. They had a guy by the name of Louie who filed the saws every night. If you and I were partners, he knew how we liked our saw. The fallers picked up their saw every morning. Then they tried to introduce power saws.*

*I remember Mark Snobenek saying, "By Jesus, I'm never gonna have a damn power saw in my camp!"*

*"Mark," I said. "Come on now. You can't be serious. Once these guys learn how to use power saws, they'll never go back to a hand saw."*

*"By God, not in my camp. I won't hire anybody that. . ." A year later the whole dang camp had power saws.*

*Well! The company made the mistake of trying to furnish power saws to these guys, and they would bring the saws home in a basket. And Louie didn't know how to fix them. Until they changed the policy and said, "You want to cut wood for us? Fine, you buy your own damn chain saw," it didn't work.*

Tom Walbridge

During the early post-war period a couple of other developments in mechanized falling and log preparation created opportunities for new equipment. One was a tree shear developed in 1954 by Tom Busch at International Paper's Olla, Louisiana, operations. A decade earlier the Mammoth Tree Shear Company in San Antonio, Texas, had marketed a shear which mounted on the front of a crawler tractor. Its scissor-like blades were activated when the tractor pushed it against a tree. Busch's shear was hydraulically powered and used on a farm tractor. Neither of these machines found much acceptance, but they were the beginning of an entirely new approach to timber falling.

The second line of development was a variety of fixed saw, called a "slasher", used to cut trees into pulp wood or, less frequently, sawlogs. The simplest of these machines — such as the one sold by Link Brothers Welding and Machine Shop in Ocala, Florida, for example — consisted of a large circular saw mounted on the front end of a farm tractor and powered by its engine. The tractor was driven up to the limbed tree and the saw cut through it. Other models required the trees to be lifted so that logs could be cut from them. The Link machine bucked up to forty cords of pulp wood a day.
The Montague slasher, which appeared about 1950, consisted of a cart mounted on a short set of tracks. A log was placed on the cart and fed into a 75-centimeter circular saw blade mounted on a wheeled trailer and powered with an air-cooled gas engine. As they were bucked off, the logs dropped onto a conveyer which loaded them into a pile or onto a truck — up to sixty cords a day of bucked and loaded wood. A self-powered, mobile slasher, the Nesco Slashmobile, was built in northern Ontario for the Marathon company in the early 1950s. It pulled logs from a cold deck with a winch and cable and dropped them onto a set of feed rollers which advanced them into a one-meter circular saw. The bucked pieces dropped onto a conveyer which loaded them on a truck at the rate of up to 190 cords a day.

The Korean War, which sparked an economic boom in Western nations, had a substantial impact on the development of forest machinery. In Europe the boom contributed to relief of the dollar shortage and elimination of import restrictions. World trade flourished. The boom caused by the war and the deregulation of imports are now seen as the catalysts for a rapid expansion of the forest industry in Sweden and Finland. It was during the 1950s that the mechanization of timber harvesting began in northern Europe.

During the late 1940s, European forest workers began the laborious task of rebuilding their industries through a combination of ingenuity and salvaged equipment. A good example of this process was the use of wood-gas generators to fuel logging trucks. These generators, mounted on the sides of trucks, converted wood chips to a combustible gas. It was an inefficient process, requiring constant stoking which left the drivers covered with soot after a day's work. In Finland, burdened with war indemnities to the Soviet Union, a fleet of thirty trucks — mostly American models, including numerous war-surplus vehicles — was assembled at Koirajärvi for the 1948–49 winter logging season to haul logs over a twenty-kilometer ice road through the swamps. Sweden, where metal industries had prospered during the war, was in much better condition to undertake a post-war mechanization program, and immediately after the war began importing crawler tractors from the US and western Europe.

The Korean War and the economic activity it fostered raised stumpage prices and logging wages, provoking a need for mechanization. During the 1940s, only thirty chain saws had been imported into Finland; by 1952, another five hundred had been purchased by loggers. It was a

"

*With the dealer-producer system you came in my office and sat down and you told me that you lived in Wilcox County and you knew a lot of people, and you knew a lot of land owners, and you wanted a wood dealership. And I said I'll sign a contract with you that we will buy from you 40,000 cords of wood a year delivered on rail cars at our plant at Mobile. I'll pay you — in those days — $25 a cord for the wood as delivered. We sign a contract, you left, and I didn't give a shit who you bought the stumpage from. I didn't give a shit how you got it on the rail cars. I didn't give a shit where it came from.*

*When a car got down there to the mill, my guys went down and scaled it and we wrote you a check for $25 a cord for every cord that was on that car. You dealt with people that were called subcontractors, or producers. And generally they were little bitty guys and a lot of 'em were a guy, his son-in-law, and his son. It was called the dealer-producer system and it depended for its success upon having an almost unlimited supply of marginal labor, principally black, who would work their ass off because some white man wanted 'em too.*

*It was mechanized a bit. You didn't gnaw the tree down with your teeth. When I first got in the business in 1950, you had a three-man crew. They might have a mule out there. They drove the truck to the stump, cut the tree down, loaded the truck, and the whole damn crew got on the truck and drove into a rail siding and loaded the rail car from the truck by hand.*

*The first thing, really, that came into being was — this was about 1953 — was the sling loader in the wood yards. This was a remarkable breakthrough because you didn't have to stop cutting in the woods to load the rail car. A lot of companies then begin to have wood yards of their own, a lot of dealers had 'em. They unloaded the truck and put it on the rail car. While the truck driver was going and coming, the rest of the crew could stay out there and cut wood. It really picked things up.*

TOM KELLY

period of great enthusiasm among machine builders. Everyone in Sweden seemed to be developing logging machines, each more sophisticated than the next.

A forwarder Calle Larsson, a Swedish logger, built from a cut-up Volvo truck might have been the spark which ignited the blaze of inventions which followed. It was built in the winter of 1951–52 for Kopparfors AB, based on an idea borrowed from war-time half-track trucks. Larsson combined the front end of an 85-horsepower Volvo with a set of half-tracks on the rear. A sleigh was hooked on behind. Larsson, hauling logs with his machine in the Grannäs forest, made his return trips across the frozen lake at eighty kilometers an hour. Word of this feat spread and soon Ösa, which built wagons and sleighs for logging, showed an interest, hiring Larsson as head builder — an event that marked the beginning of modern Scandinavian forest-machinery development. The forest companies began to actively participate in the development of new machines. For example, Bergslaget's forest manager, Bo Hedegård, began spending one day a week at the Ösa factory, helping to design new logging machines.

Farm tractors had become common in Sweden. The next phase was to adapt them to logging by adding half-tracks. Volvo, which acquired Bohlinder-Munktell's (BM) tractor factory in 1950, asked Ösa to outfit its BM-10 tractor with half-tracks. The machine was displayed in 1952 at the Eskisltuna agricultural exhibition where it attracted a great deal of attention and encouraged other manufacturers to adapt other makes of tractors for logging. By the end of the first year, Ösa had adapted one hundred farm tractors. Half-track tractors were soon in use as carriers for excavators, ore prospecting tractors and stone-clearing machines. Half-tracks were also installed on rear drags, enabling logs to be skidded in the summer, ending the reliance on winter logging. Forest companies understood immediately the advantages of year-round logging. Loggers were freed from the limitations on access provided by river channels and engineers were able to build road networks and railway lines wherever they were needed.

From 1954 to 1959, snow trials were held for the new logging machines in a mountainous area of Sweden where dozens of prototypes were sent for testing. Included among them were a Swiss-built snowmobile constructed from a Willys jeep, and a Bombardier snowmobile from Canada, christened the Snow Comet, that raced across the snow at sixty kilometers an hour. None of these machines were well suited for logging, but no one had seen anything like them in Sweden so everyone wanted to try them out. One of them was the first Swedish logging machine on wheels. It was a forwarder designed by a Garpenberg Forestry College engineer named Troeng and built from a Svedala road grader. But the eight-ton machine was too clumsy and heavy for forest work. Most of the machines tested could not handle deep snow, and the trials often amused observers who watched as they slid around and sank into the snow. While few of the machines, imported or domestic, did well in the tests, they provided Sweden's foresters with an opportunity to see what was being built in other parts of the world and to evaluate where developments were heading. The poor showing of foreign machines encouraged Swedish designers to continue their own work.

The Swedes also tried out conventional crawler tractors, most of them imported from the US. Again, they proved unsuitable for working in the forest; they were too heavy, costly and overpowered for the task. In 1942, Sweden's KF cooperative, the country's largest farm cooperative, imported a US-made Cletrac crawler, a popular North American small log skidding machine introduced by the White Motor Company in 1918. In 1944, the Cletrac company was taken over by the Oliver Corporation, and beginning in the early 1950s, Ösa imported and modified about 700 Oliver Cletrac crawlers. In 1954, Ösa developed a new drive system for crawlers, called Alfta bands, utilizing rubber tires instead of sprockets inside the tracks. These were mounted on the Cletracs, as well as a variety of imported crawler tractors, which were used by loggers for building roads, scarifying and, in a minor way, skidding logs. The modification of existing machinery by the addition of Alfta bands was a major advance in the mechanization of Scandinavian logging.

To this point, all the mechanized equipment used in Scandinavian forests, apart from a few chain saws, had been imported and, occasionally, adapted to local needs. In 1956, the first Nordic-built machine, the Bamse crawler tractor, was introduced through a joint effort of Volvo, BM and Ösa. The Bamse had a two-cylinder, thirty-horsepower engine, and came equipped with a dozer blade and a winch for skidding logs. A later version included a loading crane which could be operated by a remote control, allowing the operator to both skid logs to roadside with the winch and load them onto a sleigh or trailer. The compact, highly maneuverable tractor was an immediate success and more than 800 of them were built over the next few years, with sales all over the world. It marked the beginning of a forty-year Scandinavian drive for supremacy in the forest equipment business.

Meanwhile, in 1951, the Rauma-Raahe forest company in Finland imported an American Allis-Chalmers crawler tractor to its northern Oulu operation. Loggers dubbed it the "American Alli" as it set about proving its worth. It cleared a skid road for itself, then pulled three-hundred-log loads to roadside. A second American machine, an army surplus amphibious cargo carrier, was put to work as a service vehicle and to pack snow for haul-road bases. Rumors that several tractor logging sites had been established in Sweden reached northern Finland that year, and crews were sent to observe them. As a result, the Oulu operation acquired a British Fordson Major farm tractor, for which a sleigh was built.

"

*Before the Second World War, wood generally was brought in by people that farmed, so it was very cyclical. You had to have high inventory for the winter and this sort of thing. They would put wood in after their crops were in, and they would put wood in after the crops were out. It was predominantly farm labor in old beat up trucks. Then the war came along and, of course, they had a helluva time. I think they actually used German prisoners at some of the places in the South, and also in Michigan, for labor. But since it was all short wood you cut it to 5-foot lengths because it was so much heavier than spruce. You couldn't cut it to 8-foot because you couldn't manage it. It wasn't too much trouble to take two or three men out in the woods, hand load the truck, and drive it out of the woods.*

*The industrialization of the South after the war made a difference in that people moved into the carpet mills and moved into the textile mills and said to hell with going into the woods, working in hundred and ten degree heat and a hundred percent humidity, when they could work in a mill that was air-conditioned and steady work.*

*Those who still wanted to stay in the pulpwood business did. I can remember when you could buy a bobtail truck for $200. Get a chain saw, hell, you could go into business as a logger. Wasn't any problem if you could get somebody to buy your wood.*

Tom Walbridge

> *In the early 1950s, we were developing forest machines from Ferguson farm tractors. We installed long caterpillar treads and caterpillar tread steering. The first tests did not go very well. We had difficulty getting the steering power and the long treads to work in balance. The machine dug itself deep into the snow in turns. We worked on it some more, and soon the new version was ready for the snow test. When the test driver returned from the test drive, I made my way through the deep snow to meet him and shouted: "So, how did it go?" I received an irritated and terse answer. "I'm sorry Carl Einar, but could I give you a good piece of advice? Don't show this to anyone!"*
>
> CARL EINAR MALMBERG

Finnish farmers were beginning to buy tractors for use in their fields, so there were plenty of willing timber-skidding operators available, especially in the winter. But the tractors they possessed, as well as their driving skills, were questionable. For example, the front wheels of the tractors tended to rear up when starting off with a turn of logs, a problem that was solved by installing a limiter — a long pipe attached to the back of the tractor which pressed down onto the sleigh when the front wheels raised. By the mid-1950s, a variety of crawler and wheeled tractors were used for such tasks as digging drainage ditches and scarifying. Several Soviet-made TDT-40 tracked skidders were also tried but, like the Soviet army, they too had to surrender in the middle of the Finnish forests. The tractors were plagued by breakdowns, and even on prepared skidding roads their performance was mediocre.

Elsewhere in Europe, other attempts were being made to transform farm tractors into logging machines. To an even greater extent than in Scandinavia there was need for small, compact logging tractors to work in relatively small forest holdings, many of which were managed in conjunction with agricultural operations. The Swiss-made Plumett farm tractor was equipped with a set of winches enabling it to cable yard logs. As well, the power take-off shaft was adapted to drive a live axle on a log trailer. An all-purpose work machine for farms, the British-built Straussler Lypsoid, was a twin-engine tractor equipped with air-cushion tires and a winch. Designed for carrying five-ton loads over rough or swampy ground, it was tried as a log skidder with little success. Another British company, Packman Machinery, developed a self-propelled powered sulky consisting of an arch and an air-cooled engine mounted on a set of two wheels and an axle. A log was raised into the arch at its centre of gravity with a hand winch. The engine drove the wheels and the operator steered by guiding one end of the log along a forest trail. Road Machines Limited, a Middlesex manufacturer, produced a monorail log transportation system made up of a portable track upon which a carriage equipped with log bunks was pulled by a winch and cable.

A small, ten-horsepower, six-wheeled, frame-steered forwarder, the Clowez, was built in Fourmies, France. Up to two cubic meters of logs were loaded crosswise in a bunk in front of the engine, and lengthwise on a trailer at the rear. The driver sat above the engine between the loads, looking like the driver of a scooter equipped with a giant shopping basket filled with logs. The French also attempted to adapt farm tractors for log skidding, one of the most successful being the Agrip, produced by the Parisian shop of Somater 56. It came equipped with a winch, a ground anchor, and a choice of an Aster or Berliet engine. With four-wheel drive and an oscillating front axle, the Agrip operated on slopes of up to forty percent. A variety of half-tracks, similar in concept to Alfta bands, were produced in several European countries for use on farm tractors. The Austrian-built Motormule, which came in both wheeled and tracked versions, had a front-mounted winch and was used to skid logs and pull a trailer loaded with logs.

Despite their dependence on imported machinery, the Finns produced their own unique machine during their initial foray into the logging equipment business. In 1953, the world's first prototype delimbing machine, Oksa-Olli, was built in Finland. Weighing only 250 kilograms and equipped with a three-horsepower engine, it was suitable only for trees under eleven centimeters in diameter.

In North America, where the mechanization of tree falling and log preparation was, in hand, primarily through the use of chain saws, the next phase of the operation required machines to replace the horses and mules still used everywhere except the West Coast. In pulpwood country, standard practice, even with the use of power saws, was for fallers to buck the wood into short lengths and stack it. There are several explanations for the evolution of short-wood logging: it was a log size a man could lift; a truck or rail car could accommodate a pile of short logs loaded by hand from each side; and cutting logs into short lengths prevented their diversion to sawmills. Whatever the reason, it had become so much a part of the logging process in every pulp region, and so much of the existing technology in the woods and at the pulp mills was built around it, that short-wood logging prevailed for decades. In 1951, for example, 95 percent of the pulpwood produced in eastern Canada was handled in this manner.

At the transfer points and mill yards, cranes and other machines began to appear to reload short logs, but in the woods, until well into the 1960s, most of this work was done by hand. It was the hardest part of a wood cutter's job, and a big part of the reason it was hard to find good people to do the job in the booming post-war economy. As a result, one of the first challenges for the industry was to come up with a machine to move piles or bundles of short-log pulpwood from where they had been cut to roadside. In eastern Canada one of the earliest solutions involved the use of self-loading sloops: sleds towed behind tractors with a loading arm that tilted a one-cord pile onto the sloop, which could then be pulled to the road. The first Canadian company to manufacture these devices was a new firm in Woodstock, Ontario, with some deep roots in the West Coast logging industry. Timberland Machines Ltd. was started in 1947 by Bob Symons and Wes McGill. Symons was a Vancouver Island logger, a self-taught mechanical genius who had gone to work for Frank Lawrence before the war building cable yarding systems. In 1945, he travelled east with Lawrence to demonstrate a Model 10-10 cable yarder to a CPPA equipment show in Quebec. A short time later, he set up a sales outlet for Lawrence in Ontario to sell eastern loggers the kind of cable equipment used on the West Coast. Symons soon hooked up with McGill, a salesman for a Woodstock trailer manufacturer, who provided initial financing. They formed Timberland Machines to build logging equipment and, finding limited interest in cable yarding machines, began producing whatever logging equipment they could sell. One of their first products was the sloop.

Within a few years several attachments for a variety of existing machines were developed to load and carry pulp wood. The Hyster company of Portland, Oregon, came out with a modified arch that carried four cords of pulp wood. Bombardier, a Quebec snowmobile builder, developed the HDW to carry 1.2 meter pulp logs and met with good market success. Radisson Tractor and Go-Tract Systems, two other Quebec companies, and Nodwell, a Calgary, Alberta, rubber-tracked-vehicle builder, all came out with tracked forwarders for moving pulp wood out of the woods. Another device with a similar purpose was the Nelson frame, which was mounted on the rear of a crawler tractor. The winch line hoisted a one-cord bundle of wood onto the frame, and the tractor carried it to roadside. A similar device, known as a lifting fork, was developed in

“

*About the time chain saws first came in, around 1952 or 1953, they had Farm Week at Michigan State University. I went to that and these guys was there from the upper peninsula selling Titan chain saws. I bought the first ones in lower Michigan and they made me a dealer. I didn't want the dealership so I gave it to my brother and he took care of our saws then and sold some on the side. We would use his saws for maybe two or three months and he could sell them as used ones for pretty near as much as he paid for them. So he kept us in good saws all the time.*

*Then Disston come out with a two-man saw. It was so heavy it killed the one guy hanging onto it and just about the other guy too. But we did run them for a little while. And then they come out with this here Homelite one-man saw. That kind of started everyone into the saw business.*

NORVAL MOREY

France for use on farm tractors. The Johnson forwarder consisted of a grapple with a one-cord capacity mounted on the rear of a crawler tractor. It operated in much the same fashion as the Anderson Woodhawk, a wheeled or tracked trailer developed and used in the U.S. Lake states. It too used a large grapple to grab the bundles, lift them off the ground and haul them to roadside.

All these approaches consisted of adaptations of existing tractors or tractor accessories. While many of them speeded up the process of forwarding pulp logs to roadside, they were still frustratingly slow, and too expensive to find widespread acceptance. What was needed was a machine designed and built specifically for the purpose. The CPPA Woodlands Section assigned itself this task in 1950. Its work, funded by a group of thirteen pulp and paper companies, resulted in a series of prototype models built by Bonnard Equipment Limited of Lachine, Quebec. The first model consisted of little more than a self-loading trailer pulled by a farm tractor. But it led the way to the Mark II Forwarder, a machine that incorporated a radical design feature — articulated steering. Also known as frame, bent-frame or trunnion steering, this system employs a pivot point near the center of the machine. The front and rear sections of the machine ride on fixed drive axles and steering is controlled by hydraulic rams which bend the machine around the pivot point. As a result, the rear wheels of the vehicle follow exactly in the tracks of the front wheel, making it very easy to steer over rough ground and providing a high degree of maneuverability. As a bonus, the articulated steering action provides added traction.

Essentially, the Mark II was a farm tractor with its front wheels removed and an additional set of drive wheels attached to the rear with the steering trunnion. With some improvements to its loading mechanism, production models of the machine were sold in 1954 as the Bonnard Logger. The machine was prone to breakdowns in the woods, however, and eventually the sponsors withdrew support for further development. In retrospect, the Bonnard series was a landmark development, paving the way for an entire new generation of logging technology, which almost fifty years later still dominates the field. At the time, however, there was a great deal of scepticism about the forwarder. A year after the Logger was unveiled, the relationship between the Woodlands Section and Bonnard broke down, and rights to the design were sold to Clark Equipment of Buchanan, Michigan, which mounted a Carco winch on the rear and began selling the machine in 1958 as the Michigan PL-75 skidder.

In the southern pine forests of the US a parallel development of prehaulers or forwarders occurred. Throughout the 1950s in the South, the most common skidder was still a mule. A good logging mule cost $150 to $250, it could work on the farm and skid logs year round, and it was, until well into the next decade, still the cheapest way to move wood from the

“

*What agriculture was left — there was still some cotton and tobacco — relied heavily on seasonal labor across much of the South. With tobacco, you needed labor extensively at transplant time and when you harvested the leaves. There was a surplus of agricultural labor throughout much of the year. So what you would do is move that labor into the woods and they would become your pulp-wooding crews. When I first came south, it was tobacco time and there wasn't a damn stick of wood to move. And so pulp wooding mixed very well with the excess agricultural labor. And starting in the late '50s, we started getting agricultural chemicals perfected and agricultural mechanization where farmers didn't have to maintain this large pool of tenant farms. And as a consequence, that agricultural pool started to drop. We had the social changes associated with the civil rights movement and the industry had to mechanize to compensate for that. And the thing that triggered mechanization down here was agricultural chemicals and agricultural mechanization.*

*At that time, man-day productivity with a guy with a truck, a chain saw, three men — a good crew could get about 1.8 cords per man-day. Because they had to cut the tree down, they'd drive the truck to the stump most of the time, cut it into 5 foot 3 inch pieces, tumble it out to the truck road and then hand load it. And, with the industry expansion that was taking place, if you took the number of cords industry needed and divided by 1.8, you soon discovered everybody south of New York City and east of Dallas had to be out here cutting 5 foot 3 inch wood, and we knew that wasn't going to happen.*

*Because the timber was privately owned and the companies were expending their capital on their mills, we had a situation where we needed to mechanize but we had no external source of money to pour in. The system had to develop its own capital. And so what we ended up with was an outgrowth of industrial and military equipment at the end of World War Two. And starting out with some relatively primitive machines.*

Bill Stuart

"

*The South was very similar to the Appalachians. The traditional methods of logging broke, for the most part. Once the first harvest was completed, the industry imploded, went back to a cottage industry. Then as timber recovered, a new industry came in. So we didn't carry across a great many of the traditions from the days of the big mills and railroads and so forth. The ties had rotted, the steel had gotten rusty, and the pistons in the old steam engines had seized. So when the next crop came in, a new technology had to come in. But there were still tram roads and mill sites and so forth that we could use. And the foresters who were involved in harvesting the second and the third forest, had the history of the first cut written in the ground, and when you went in, you learned. You went in and surveyed what the old folks had done because you could use the same railroad grades, you could use the mill sites. It saved you a helluva lot of work.*

*And many times you could go back in and still tap some of the old roads, the old horse roads. The old bobsled roads were adequate for a D4 and an arch to get up. Because that was one of the things that made the D2 and the D4 so nice, they had about the same width track as a team of horses or a span of oxen. So you could use many of the old roads.*

*The cable systems they were using in the west were too expensive. And in that period, after it imploded to a cottage industry, it was damn hard for a cottage industry to raise capital. So you had to raise capital one buck at a time. . .*

*The way it worked, many of the mills would come in and they'd put a car on the siding and they'd pay a store owner, or contract with somebody, "Well, I'll pay you X dollars if you have that car loaded in a week."*

*Well, he'd get in touch with people and say, "I got orders for pulp wood," and folks'd take the fertilizer truck, go out and cut 5 foot 3 inch pulp wood and load it on a rail car. They'd send an engine in to switch that car out and take it to the mill. And our dealer system grew out of that because so much of the timber was in private ownership. You had a local individual then who was the coordinator of wood supply out of this depot town. He'd say "I got the truck. I'll sell it to you for so much a cord." And all you had to do was bring the chain saw and sweat, and you would become a logger. And there was a helluva lot of wood poured out in the late '40s and '50s, with that type of a system.*

Bill Stuart

stump to roadside. What was missing in the equation were men to work the mules. They preferred to work in an air-conditioned factory for steady wages instead of following a mule through a southern pine forest in 95 degree heat, with humidity to match.

Throughout the South at this time, the most widely-used mechanized logging system consisted of a bow or wheel saw to fall trees and buck them and a bobtail truck to drive into the woods and haul the logs to the nearest wood yard. The truck was most likely a 1.5-ton-rated, single-axle vehicle that was loaded by hand when employed hauling pulp wood. There was an unending supply of used trucks available: old Coca Cola trucks which had never been off the highway and could be picked up for $500. In the southern US, a man could become a mechanized logger with an outlay of a thousand dollars or less. But this primitive system could not supply the rapidly expanding pulp-wood market, as many innovative machine makers realized. One of the first additions to the basic setup were pallets, particularly the Dixie Pallet System developed by Tidewater Equipment in the late 1940s. These were nothing more than bent pipe frames on skids which held a cord of hand-loaded, short-log pulp wood. Once loaded, the pallets were skidded behind a crawler tractor to roadside, where two or three of them were winched up a tilting frame onto the truck and hauled away. They were reputed to increase a crew's output by one-third to one-half.

In 1954, John Thibodeau, a Quebec mechanic, and Carl Brice, a pulpwood dealer from Archer, Florida, introduced a small four-wheel drive, winch-equipped tractor, the Go Getter, designed to skid logs or short-haul palletized pulp wood from stump to roadside over the often muddy southern coastal plain. A distinctive feature of the machine was a set of hydraulic cylinders between the frame and front axle which provided front-end oscillation to aid in traversing rough ground, and also allowed the operator to control weight distribution to improve traction. The machine was not a success and, like many others of this era, soon faded from sight.

A similar pre-haul tractor came on the market in the mid-1950s, the Harrison Pulpwood Harvester, built in Columbia, South Carolina. Like the Bonnard forwarder, it used a farm tractor with its front wheels removed, and a trailer attached to the rear with an articulating steering trunnion. It winched pallets onto its tilting trailer, or loaded itself with a hydraulic loading boom. Its great weakness was a lack of power to the rear axle, and it proved more or less useless off the road. Walter Jarck, one of the country's most inventive logging equipment engineers, who worked for Catawba Timber in South Carolina, came out with an improved version. Jarck added another drive shaft to power the rear axle, producing a successful forwarder he also called a Go Getter. He then made an unorthodox move for a design engineer. Instead of developing his machine into a commercial product, he printed up copies of its design plans and sold them to all comers for thirty cents each. More than 300 sets were distributed. With these plans and some army surplus truck parts, a local machine shop could build a forwarder for a few thousand dollars.

"

*McCulloch started in Los Angeles, I think right at the end of the war. The oldest McCulloch I've got is a model 1225 and that would have been made in 1947. He was very innovative. He came in with die cast castings when everyone else was using sand cast castings. That makes a big difference in weight. Beavertail bars was another one of his ideas. He made his own carburetors. Everybody had been using Tillotsons prior to that. Quite a few little things. They were the king of the heap when I started in '51; there were more McCullochs and IELs around than anything else.*

*They were hard to work on. Jesus, they were hard to work on. The starter housings, the bolts going in the starter housing, they had captive nuts behind them. Instead of having threaded castings they put a nut in a little socket and if you ever dropped that out of there, of course, you couldn't tighten the goddamn screw. You'd have to take everything off and try again. It was a nightmare.*

*The carburetors were not as good as they should have been and the ignitions got damp. They were always a problem. But they were really good engineers those McCulloch guys because they developed things other guys just hadn't done at all.*

Dave Challenger

The new concept in steering was introduced to Sweden when Volvo-BM started manufacturing framesteering, front-end loaders and dumpers for the mining and construction industries. Inevitably, the idea was applied to forest machines. In 1954, Calle Larsson produced the first Scandinavian frame-steered logging machine. He shortened a truck and attached a trailer behind it, with a steering trunnion in the middle. The initial test drive did not go well; the machine set off down a country road and downed dozens of mailboxes, after which it was known as the Moon Rocket. Bergslaget's Bo Hedegård designed a six-wheeled machine, called Lotta, with the front end of a truck, frame steering and a truck's trailing bogies. In addition to the frame steering, the front wheels also steered. Lotta was not a success, but it ushered in another major innovation. The Swedish hydraulics company, Hiab, brought onto the market a new type of hydraulic truck loader, several of which were mounted on Lottas. It was a radical new concept which eventually became the dominant log-loading system around the world. Improved hydraulics, a key component of frame steering and the new loading system, was at the heart of the revolutionary changes in logging equipment design which began to appear in the late 1950s.

Falling hardwood trees in the southern United States with a Lowther wheel saw. These saws were used extensivley on level ground before the development of suitable chain saws.

Bucking pine pulpwood with Kut-Kwick's Knight wheel saw in Georgia, US, in 1941. After the tree was felled, it was delimbed with axes before it was bucked.

Bucking tropical hardwood logs in Liberia in the 1960s with a chain saw.

Loggers operating a two-man bow saw in the southern US in 1953.

In the 1950s, tractors replaced horses on Scandinavian farms. Farm tractors were dangerous to drive over difficult terrain, especially when skidding or hauling logs.

The BM-10 farm tractor, the first in Scandinavia to be equipped with a half-tracks invented in 1952.

The 1950s in Sweden is called the era of half-tracks. All tractor manufacturers and importers equipped farm tractors with half-tracks for logging. The log sled used on this model was held over from the age of horses.

Half-tracks became almost a fad. They were installed on excavators, earthmovers, combine harvesters, and even the mining exploration tractor pictured.

Loading pulpwood the hard way — by hand in the southern US in the early 1960s.

Swedish loggers admiring the ultimate achievement of the half-track era, the BM 400 forest tractor made by Ösa. It was equipped with a cable loader and a trailer suitable for summer use.

A Harrison Pulpwood Harvester, working in Georgia in the mid-1950s, was a two-wheel drive machine designed to haul short logs from the stump to roadside. It was an early version of a North American forwarder.

The Jarck Go-Getter, designed by Walter Jarck, at work in South Carolina, US, in the mid-1950s. Scores of these pulpwood forwarders were built from plans Jarck sold for thirty cents each.

Debarking machines appeared in Scandinavian forests in the mid 1950s. The number one brand, shown working in Sweden, was the Cambio.

As more highly mechanized Scandinavian debarkers emerged, a hydraulic loader both fed the machine and stacked peeled logs.

A grader for winter roads developed by Swedish machine builder Carl Einar Malmberg. It was attached to the front end of a farm tractor, which was also used to haul logs.

Loading a pulpwood pallet onto a Jarck Go Getter. One of the first successful American forwarders, this machine had powered rear wheels.

Ösa's first track-equipped trailer, built in 1952. It tilted to the side to unload.

The Bamse, the first Swedish crawler tractor, was introduced in 1956, and was the forerunner of all Scandinavian forest tractors. It was a tremendous success, and was used for a variety of tasks, including the towing of log sleds.

The Bamse was under constant development. During the second half of the 1950s, it was equipped with a radio-controlled winch that skidded the logs close to a mechanical loader.

The reliable and stable Bamse was also used for soil preparation.

The hydraulic knuckle boom and loading grapple marked a revolution. There was efficiency in loading. Tractors saw more power, and trailers started resembling the modern loading bed.

The Bonnard Logger, working in Ontario, Canada, in the late 1950s, was the first frame-steered forest machine built. This model, built as a pulpwood forwarder, was the precursor of a long line of skidders and forwarders.

This shop-built tracked pallet cart, loaded and towed by a crawler tractor, was used in the US South during the late 1950s. The loaded pallets were skidded onto the cart's tilting bed.

The hard luck Lotta was too heavy for forests. However, it was one of the first forest tractors in Scandinavia equipped with frame steering. Designed by Bo Hedegård, the Lotta had front wheels that also turned. Additional accessories included a Hiab hydraulic knuckle boom, which was a giant step in the name of progress.

# The Rubber Tire Revolution Begins

In the 1950s an intense debate, which is still not resolved, began among loggers, engineers and the builders of forest harvesting equipment. Pared down to its essence, the discussion focused on the question: where should a tree be processed — at the stump or at roadside? Over the intervening years there have been variations on this fundamental question — some solutions divided the task between the two locations — but essentially that was the issue. It was not an academic question, but one which had immediate bearing on the kinds of machines that were being developed. For the next twenty or thirty years, the discussion swung one way or the other, as new types of machines, or a new component for existing machinery, appeared. It was a debate which occurred with varied intensity in every forest country around the world.

Discussion of this issue began in earnest in North America in the early 1950s with the appearance of the first rubber-tired machines designed to skid long, or tree-length logs. Before this time, attempts had been made to utilize existing rubber-tired vehicles. The attraction was the speed with which they could travel compared to the tracked machines then in use, offering the possibility of longer skidding distances and faster skidding cycles. Farm tractors were tried, but failed. After the war a large number of surplus six-wheel-drive military trucks came on the market and loggers and mechanics added arches, winches and other devices to skid logs over unroaded ground. They were not very maneuverable, however, and were used primarily for skidding along prepared roads.

Loggers and equipment people debate who produced the first skidder. There are almost as many claimants to the title as there are manufacturers. Part of the difficulty lies in the disagreement over exactly what qualifies as a skidder. It is a debate which will probably never be resolved. In any case, in 1951, after five years of research, Paul Westfall, a Portland, Oregon, equipment designer began marketing a rubber-tired skidder he called the Performer. It was powered with a 225-horsepower Cummins diesel, had wide tires for flotation, a dozer blade, an integral Carco logging winch and steered with the use of air-controlled steering clutches, on the same principle as a crawler tractor. The machine was built of components familiar to loggers, and it was fast and tough — a logger's skidder. The Performer operated by hooking a choker to the log or tree, winching it up to the rear of the machine, where it was raised under the arch and skidded to roadside with the tail end of the log dragging on the ground.

In the same year, the ubiquitous Bob LeTourneau began selling his Tournaskidder. It was a four-wheel drive, rubber-tired tractor with a winch used with a towed arch to skid logs. Not many were sold, probably due to the machine's high cost and its multitude of unusual technological features. Loggers who did use one — such as the pioneering West Coast skidder loggers, the Deering brothers — had nothing but praise for its ability to skid big turns over distances of three kilometers or more at high speeds. Two years later, in 1953, LeTourneau introduced the Tournarch, which utilized a two-wheeled prime mover he had called the Tournapull when it was first produced before the war. The new version was attached to a two-wheeled arch through an articulated steering mechanism consisting of an electric motor geared to a kingpin. Only the front set of wheels were powered. A year later this machine evolved into the Electric Logger, on which all four wheels were equipped with electric motors which drove, braked and steered the machine. Powered by a 235-horsepower Buda diesel, the Electric Logger skidded a large turn of logs at twenty kilometers per hour. It, too, was rarely used — only nine were sold in Canada — but its admirers, such as Viv Williams, a Queen Charlotte Islands spruce logger, thought it was one of the best logging machines ever built.

In 1952, KVP, an Espanola, Ontario, timber company, and Archie Kerr, the local dealer for the four-wheel drive FWD Truck Company of Wisconsin, stripped down a FWD truck and added a winch and A-frame at the rear. This machine was called the Blue Ox, after Paul Bunyan's legendary companion, and sold widely under that name for almost ten years. Although its usefulness was limited by its conventional, front-wheel steering, this skidder was instrumental in proving the worth of off-road, rubber-tired skidding machines.

In 1954, a West Coast manufacturer of front-end loaders, Wagner Tractor, came out with a skidder, available in several models, called the Loggermobile. A unique feature of this machine was its articulated steering, essentially the same as that used in the Bonnard Mark II Forwarder which had appeared three years earlier. Like the Performer, the Loggermobile was a ruggedly constructed machine with a 200-horsepower diesel engine, an integral arch, dozer blade, and four-wheel planetary drive. However, it was a difficult machine to keep running and the hard-surface tires used on it provided poor flotation.

In 1956, Timberland Machines in Woodstock introduced a conventionally-steered skidder, the Timberskidder, the first of a long line of rubber-tired skidders which eventually became the most popular in the world. The Ontario company established almost a decade earlier to build cable logging machines, had met with limited success and had branched out into other products. The Timberskidder was designed to go to the stump, yard in and skid to roadside a three-cord turn of tree-length logs. It had a 200-horsepower Chrysler V-8 gas engine and was intended primarily for long-distance skidding, building on the modest success of the Blue Ox. It skidded at fifteen kilometers an hour or more, and came back empty on a good road at more than sixty kilometers per hour. Priced at $17,000 dollars, it was expensive for an era when many loggers considered mechanization to be no more than a possible option. Sales were moderate, but sufficient to continue development.

The success of the rubber-tired skidders presented Timberland's founders, Symons and McGill, with a major decision. Should they proceed to make improvements on their rigid-frame machine, or switch to the articulated frame that was responsible for the limited success of the Bonnard Logger and the Westfall Performer? Fortunately for them, and the logging industry, they took the second route. In 1961, the company, in which a controlling interest had been acquired by the Ellicott Machine Corporation of Baltimore, Maryland, came out with its first articulated skidder, the Timberjack. The name had several associations. The common designation of a West Coast steam yarder was "donkey", of which males were called jacks. Male mules, used extensively in the South to skid logs were called "jackasses". The word "jack" is also applied to a variety of devices which save human labor. Within a couple of decades the Timberjack became the most widely recognized piece of logging equipment in the world.

The new version of the Timberland machine was a smaller, more maneuverable machine than the original Timberskidder, designed to replace the horse. Known as the 200 Series skidder, it had a smaller, sixty-horsepower Ford gas or diesel engine, a relatively small 2,700-kilogram winch, and sold for under $10,000. The Timberjack was an immediate, overwhelming success. In the year it was introduced, 44 machines were sold; 209 were sold the second year, 567 the third, and 828 the year following. Sales continued to climb rapidly for the rest of the 1960s, including exports of almost 1,500 skidders to Europe, Asia, Africa, Australia and New Zealand. The largest single customer was Sweden, where 328 Timberjacks were purchased by 1969.

"

*Back about 1956 Leo Heikkinen invented what I consider the first hydraulic type loader. And I think that I bought either the second or third one that Leo made and mounted it on one of my trucks. We took the cable loader off it and put this hydraulic loader on it, which at that time was the state-of-the-art, and started hauling our logs and pulpwood with this machine.*

*Now the sad part was that you had to sit on the cab. There were no seats. You sat on the roof of the cab and worked the controls. You let your feet dangle in mid-air, and worked it that way. And I'll never forget it because the first load we hauled in, my brother-in-law was running the thing, loading off a dock into a railroad car. He swung the load over, the first load off the truck, and the winch somehow stuck in such a way that we busted the whole top of the loader right off and it ended up in the box car! We took it back down to Leo and he repaired it.*

PAT CRAWFORD

> *The first loader I ever saw was a Prentice loader and it was parked at a rail siding in Virginia. It was broken, the main cylinder was broken. The guy who ran the wood yard had purchased it but it was broken and he didn't have any parts for it.*
>
> *"What is that thing?" I asked him.*
>
> *"Oh, that's one of those Prentice loaders."*
>
> *"Well, hell," I said. "That'll never work. The boom isn't long enough." It had an 18-foot boom. The big thing in the woods then was the Big Stick Loader. The reason it was popular was because the trucks, the bobtail trucks with a Big Stick, could sit in one place for a longer period of time while the ground crew would drag a cable out to the log and winch it in, buck it up and load it.*
>
> *At first people didn't see what a hydraulic knuckleboom could do because they couldn't visualize bringing the wood to the truck. They visualized trying to make the boom reach out for the wood and it was only 18 feet long. So it was a mind set there. We had to break that paradigm. That's our whole history, the evolution of our logging industry is breaking paradigms, breaking the concept that something won't work. We're having the same trouble in some of our international markets. They don't yet realize that our machines can be applied in their market until we prove it to them. So we demonstrate and show them how they work.*
>
> Mal Crawford

Close behind Timberjack in popularity was yet another West Coast skidder, the Tree Farmer, introduced by Dwight Garrett of Enumclaw, Washington, in 1958. It was a small, articulated machine with a 30-horsepower gas engine, chain drive, oscillating front wheels, a lightweight frame of welded angle iron, and a Braden logging winch. Priced at less than $6,000, the Tree Farmer rivalled the Timberjack in sales. Between them they convinced the logging industry there was a future built around rubber-tired skidding tractors.

On the West Coast a different kind of post-war conversion process was under way. For the first few years, most of the effort and ingenuity went into converting steam donkeys to diesel power. To increase the mobility of these machines, some of them were taken off their timber sleds and mounted on a dozen or more big steel wheels so they could be towed from one wooden spar to another. As the donkeys became more compact, loggers began mounting them on surplus tank carriers and oversize truck trailers to get them into the higher, steeper settings. But they were still stuck with the difficult, time-consuming task of raising and rigging the wooden spars. And this job had to be done more frequently, as the area of each logging operation became smaller. A decade earlier, a railway setting might produce 100,000 cubic meters of timber; by 1960, settings yielding as little as five thousand cubic meters were common.

In 1951, two types of machines set a new course for West Coast logging. Skagit Steel & Iron Works in Seattle, which had been working on gas and diesel-powered donkeys since the 1920s, came out with the first of its SJ series of mobile spar yarders. The first of these, the SJ-8, was mounted on ten rubber tires, travelled at almost thirty kilometers an hour, and took three or four hours to set up. Over the next couple of decades, Skagit became the most prolific builder of cable yarders, advancing to a single-guyed leaning tower, the SJ-2, that could yard, then swing logs into a convenient position for loading. In 1964, the company brought out a combination yarder-loader, the SJ4R-GT.

A few days before Christmas in 1954, Wallace and Harper Baikie, two Campbell River, British Columbia, loggers, were sharing a bottle of seasonal cheer with Chuck Madill, operator of a Nanaimo machine shop. By the time the bottle was empty they had sketched out a design for a portable steel spar. It incorporated a Skagit donkey mounted on an old White logging truck chassis, along with a twenty-seven-meter steel tower that folded down for travelling. It also had hydraulically-powered guy-line winches and the whole thing took only two hours to put up. It was an immense success and, within a few years, Madill was the world's biggest manufacturer of portable steel spars. Soon, every logger from California to Alaska wanted a portable steel spar — known in the trade as "tin stumps" — and several other companies jumped into the business. Berger Engineering of Seattle introduced a Porta-Tower. It was a twenty-seven-meter spar mounted on a sled towed into position by a crawler tractor

“

*Before the hydraulic loader, when you tried to load a tree-length log you worked your rear end off to get it up on the truck, or train, or whatever you was going to load it on, because it was cabled and it wouldn't come up even. No way you could hook it up. You couldn't pick that whole tree up, so you had to haul short logs to load them with cables. We used to use slip yokes on 'em where one end of the cable would run down here, say 30 foot, and the other end would be 10 foot, but it still wouldn't come up even. And it was dangerous. People got hurt all the time with it.*

*But then the hydraulic loader come along. I know I loaded with a stiff boom loader for years. They brought me a hydraulic loader out there to try out, going to sell it to me. It sat in the woods two weeks. I never would use it 'cause it'd take too long to load a load of logs with it. And then one day my stiff boom broke down and I had to use that thing. By the time I'd used it all day they come out there to get it. I told 'em there was no way they was going to haul my hydraulic loader out of the woods. I was going to buy that thing! And they was little old bitty things. Now we got loaders that pick up a hundred thousand pounds. Back then, if you picked up a 16- foot log it was doing real good.*

Tom Burch

equipped with a double-drum winch, which then served as the yarding donkey. Frank Lawrence came out with a three-drum slackline machine, the Sparmatic, equipped with 275-horsepower Cummins diesels and featuring telescoping spars up to 36 meters tall.

There were also a few odd-ball cable machines that never really got off the ground, but which attracted a lot of interest. A few attempts were made on the West Coast to use the Swiss-made Wyssen skyline cranes, fixed skylines over two kilometers in length. The Wyssen systems were the product of European cable-logging development which centered in the mountainous countries of Switzerland and Austria during this period. With the Wyssen system, a winch at the top end lowered a carriage carrying logs which were released at the bottom. Designed for permanent or long-term installations, they were not a great success in North America, where the pattern of logging was to cut and move on.

In 1950, an Oregon logger named Phil Grabinski built a cable machine called a Skyhook, which required an operator who did not know the meaning of fear. It had a large carriage, powered by a small engine, and rubber tires for travelling on a road. It was also equipped with a set of grooved steel wheels which ran along two fixed, parallel skylines. Riding in the machine, the operator lowered a skidding line to pick up logs, which might be a hundred meters or more below him. Once they were attached, the logs were winched up to the carriage and carried to the landing. A similar radio-controlled machine, called the Skycar, was built by Skagit. Neither enjoyed brisk sales.

Beginning in the mid-1940s, a few Ontario and Quebec forest companies acquired cable systems. Frank Lawrence brought a five-drum yarding crane east in 1945, arousing a great deal of interest. By 1951, there were about ten cable yarding operations in existence in eastern Canada, a third of them yarding short-wood bundles, and the rest tree-length logs. Timberland began building four-drum yarders mounted on sleds with timber A-frames, several dozen of which they sold to Ontario and Quebec companies during the 1950s and 1960s until constraints contained in union agreements ended the use of cable systems in the region.

In the southern US at about this time there was a brief but significant revival of cable logging. By now Lidgerwood skidders and McGiffert loaders were becoming a distant memory. In the post-war period, with the development of power saws and the growing use of trucks, there was a need for improved mechanization of the skidding and loading phases. Two machines filled this void. The first was the Logger's Dream, produced by Taylor Machine Works of Louisville, Mississippi. It was composed of a one or two-drum winch and a welded-pipe A-frame mounted on the deck of a two to four-ton truck. The tongs on the mainline were often carried out to the logs with a mule, and the logs were skidded to wherever the truck was able to travel — which in southern pine country often took in a considerable area. The Logger's Dream was also a highly efficient loader.

It was a useful and affordable machine particularly adept at skidding logs out of wet spots or gullies, and was a southern favorite until it was superseded by skidders a decade or more later.

The other device, as common throughout the south as bobtail trucks and widely used in northern states as well, was the Big Stick loader. The name, which was applied generically to this type of loader, was a trademark of Brown Machinery Service in Louisville, Mississippi, builders of the original model. It was made of a short pipe mast or frame and boom mounted behind the cab of a truck. A winch operated off the truck's power takeoff. On some homemade versions, a brake was applied to one rear wheel, with the other one raised off the ground and used as a winch. More sophisticated models had a hydraulically-operated boom. Primarily a loader, the Big Stick also yarded bundles of pulpwood thirty meters or more. Several companies, most of them small operations, manufactured the loader, and basic models sold for $600 to $700. The Brown version, which cost $1200 and was known to be "as handy as a pocket on a shirt," was considered to be the most reliable. The real significance of this mechanism was that it was the first practical mechanical device to eliminate hand loading of pulp wood, one of the most backbreaking jobs in the woods. For that achievement alone it earned the everlasting gratitude of an entire generation of pulpwood loggers.

The development of loaders was a widely dispersed and diversified phenomenon in the 1950s, reflecting the distinct needs of different regions and the variety of logs each produced. In the sawlog regions of the northern U.S, a mechanized replacement was required for the horse or tractor-powered cross-haul method. One of the initial attempts utilized the power of the truck itself, like the Big Stick. For a brief period an accessory called the Timber Tosser was sold by Timberland Machines. One of the Woodstock company's first products, it

"

*I tried the first Pettibone that was built. And one of the first Timberjacks, I tried it too. There wasn't much difference between the Timberjack and the Tree Farmer then. There wasn't a problem with the skidders. It was the people that were selling them. We were using D4s and a D7 with arches in the white spruce and white pine. The kid that was selling the Timberjack up here on the north shore brought one of them little Timberjacks up and he said, "Unhook that D7 from that log!"*

*You know, a D7's a pretty good-sized tractor. He hooked on to it and she rared up. And on the way down he let the clutch out and of course tore the hub right out of the clutch and she was done. So we pulled him up on top of the hill, couple of days he had her all fixed up. I said, "Better stay out of the big stuff."*

*I had a D2 skidding cedar poles, so down the hill he went to get some of them. On those first skidders there were no guards on the side. The first thing, a dead balsam that was sticking out came in through the side, and took the distributor off! The whole shebang! Well, that was it. I said, "I got no time to screw with you guys. Get out in the goddamn farming country. Them rubber tired things, you can't use them in the woods!"*

*But I was wrong on that statement. It was just a matter of convincing guys like me. An old fellow I knew tried to sell us a whole lot of different things and he come up once with a Garrett Tree Farmer on the back of the truck. I said, "Gus, you're at least two hundred miles too far east and north. You gotta get back to the prairies with that thing."*

*"Oh, don't be so smart, Ray," he said. "I'd like you to try it."*

*"No," I said. "I was the laughing stock with that Timberjack." The old timers who worked in our camp had really laughed over that. "Well," I said, "if you dump it off on a Saturday night, most of them are gone. I'll play with it on the Sunday."*

*And sure enough, he did. I went out Sunday morning, down over the rock ledge where I'd been running D4s. True, I couldn't even hardly take a third of the load but I could make four times as many trips up and down the hill in the same time. And that's what sold me. The speed. So I bought two of them.*

RAY HAHN

> “
> *When timber companies started to build rail landings all over the South they had to have some way to load pulpwood onto rail cars rather than by hand. Taylor Machine Works modified the Logger's Dream to be used as a loader. It would go under a bundle of wood, lift it right off the trucks, waddle over, and set it down on the rail car. And I was around when that happened, and bought some of the first Logger's Dreams that had been modified. We called them Pulpwood Dreams. You could buy them for $12,500. The nice part of it was there was so many little bitty railyards where all you had was a track and you could only load from one side. They rigged it in such a way that you could load the offside of the car.*
>
> *My boss bought twelve of those damn things right off, like that. We were talking to a fellow at Taylor Machine Works. "Frank," my boss said. "I wanna buy about a dozen of these machines, how often can I get delivery?"*
>
> *"One every three weeks, sir," Frank told him. Later he said to me "Jesus, that's the biggest order I ever got in my life." He was a hero when he got back to the office.*
>
> TOM WALBRIDGE

was a major factor in establishing its name among loggers. The machine consisted of a loading arm, about the length of the truck bed, which hinged down from the side of the truck. A log rolled onto the arm was raised up by a winch working off the power takeoff and dumped onto the truck. By revving up the truck engine it was possible to load a second truck parked on the far side of the one equipped with a Timber Tosser. The major drawback to the machine was that it was hard on the truck, since the first layer of logs dropped onto the deck from a height of a meter. The frames of most trucks could not stand up to this treatment for long, and none of their lights worked because the impact of the falling logs broke the filaments in the bulbs.

The real revolution in mechanized loading began with development of the knuckleboom loader. Although there is considerable debate over who came up with the idea, its evolution in North America appears to have begun with the importation of a Swedish machine, the Hiab, in the mid-1950s. Bob Larson, an Ely, Minnesota machine-shop owner, obtained dealer rights to the Hiab and in 1958 began producing an almost identical machine he called a Hiabob.

The principle of the machine was very simple; its success relied on the recent availability of good hydraulic systems. Two short booms were linked with pivots and hydraulic cylinders. One of them was attached to a short pipe mast mounted on a truck bed. The working end of the Hiab boom was equipped with a loading tong, but one of Larson's improvements was to replace it with a hydraulic grapple. The boom swung, under hydraulic power, in a 190-degree arc. It was highly controllable and capable of precise, rapid loading, one log at a time. Larson, generally credited with launching the knuckleboom revolution, sold his company to Beloit within a few years.

At approximately the same time Larson was developing his machine, Leo Heikkinen in Prentice, Wisconsin, was refining his mechanically-operated loader by adding hydraulics. In 1958, Heikkinen came out with a knuckleboom loader mounted on a trailer that could be towed to the landing by a crawler tractor. His company, Prentice, was destined to be another major force in the equipment business. It was sold to Omark Industries in 1967.

Within a very short time, the development of the knuckleboom concept had a direct impact on the evolution of another machine, the forwarder. In 1958, the Canadian International Paper Company and Dowty Equipment of Ajax, Ontario, a British-based manufacturer with extensive hydraulic experience, set to work producing a modified version of the Bonnard forwarder. Bruce McColl, a former CPPA Woodlands Section engineer, had come up with a new concept for the forwarder that gave it more flexibility and utility. It consisted of the addition of a Hiabob knuckleboom loader. This simple marriage of machines produced a forwarder that picked up and loaded individual shortwood pulp logs with great efficiency, eliminating the need for hand piling and replacing forwarders designed to load and haul bundled wood. Known as the Dowty forwarder, this machine had a short life of its own but was the first of an entire family of machines that emerged in the following decade. In 1958 as well, a similar modification was made to the Blue Ox forwarder, a few of which were sold before that machine disappeared.

Abandoning the wooden spar and moving to light, portable steel towers created a need for a new type of loader on the West Coast. It was impractical, in most cases, to combine mobile yarders and loaders in one machine; their mobility was reduced and it was difficult, if not impossible, to yard and load at the same time. In an early innovation, the Bohemia Lumber Company of Oregon developed a gooseneck boom to mount on the new generation of shovel loaders equipped with turntables — Americans or Washingtons, usually — which used a tong on a winch line to pull in a log, heel it and swing it onto a truck or railcar. In the late 1950s, grapples, which did less damage to the logs, replaced the tongs.

In the early 1960s, Skagit came out with a rubber-tire loader, the Skagit 555, equipped with a drum and a tag line which could be attached to a tree or stump some distance away. The grapple was lowered along the line and dropped onto a log, which was then pulled in and loaded. The 555 was the precursor of the grapple yarder which came to dominate logging in parts of the West Coast — particularly British Columbia — for almost two decades.

As indicated by the early development of skidders on the West Coast, there was a widespread adoption of these machines there as well. Since the 1930s, coastal loggers had been using crawlers and arches for skidding. By the 1950s, they were looking at much smaller timber which had to come out of areas expensive to access even by truck. The ideal solution was the rubber-tire skidder, which could bring logs in from two to three kilometers away. One of the earliest entrants into this field was Caterpillar, which tried to convince loggers that the DW20, one of its conventional rubber-tired tractors used primarily on dirt-moving machines, was adaptable for skidding logs by adding an integral arch. MacMillan Bloedel, British Columbia's largest forest company, gave this rig a serious test but, along with everyone else, concluded that it was not the right machine. It was not articulated, therefore not very maneuverable off the road. While it had some utility as a long-distance skidder, it cost too much to justify buying one for this task alone.

Although the skidder was designed as a mobile winch and performed best as a replacement for horses, this did not prevent West Coast loggers from using it to skid enormous logs. The biggest problem was braking — skidders were not equipped to hold back the size of logs they were capable of skidding. A typical West Coast skidder trail in the 1950s and 1960s began in the setting and headed for salt water over the shortest route possible, often straight down a thirty or forty percent slope. After a few dozen trips down this trail, in the heavy coastal rainfall, the churning tires and dragging logs gouged out a trench five or six meters deep. Once a driver started a turn down one of these chutes, there was no stopping and no room to jump if the brakes failed. The Deering brothers once had a runaway Kenworth skidder rocket down a long cut at thirty kilometers an hour with a huge turn on behind. Before it could be stopped, the skidder raced across the flat at the bottom of the hill and into the water, where it sank from sight. The driver had to swim for his life, and all that could be seen were the ends of the logs. Another machine fished out the logs, the drowned skidder along with them.

“

*I was given the job of looking at some skidders that (Kimberly-Clark) was trying out. There was a Garrett Tree Farmer, a Timberjack and a Blue Ox. I spent two or three weeks doing some time studies and so on and so forth. The Garrett Tree Farmer was a good little machine. I think that KC bought a couple of them. And, along with the Timberjack, it was very obvious even then that frame steering was the way to go because the Blue Ox — which was a really fast machine and in a straight line could outrun anything — couldn't get around. It had no maneuverability. The best feature about them that was obvious, really obvious, to us guys that was the frame steering. Frame steering was what really turned the wheeled skidder into a skidder superior to the Blue Ox and all the other ones of that type. The Blue Ox depended on raw power and straight ahead in fifth gear you could skid logs at 50 miles an hour, but you couldn't get into the corners.*

*The thing about the Timberjack that was appealing over the Garrett at the time was the hubs, the planetary drive. The Garrett had axles like a truck, and I'm sure that's what they were, truck differentials. When it was pulling and it had a big load on it, the wheels would slip — you could see them slip. You could see that axle winding up and then letting go and, of course, once in a while it would snap off. But the other one, the planetary drive, it just turned the wheel steadily all the time. Whereas, with the Garrett, you'd see the wheel wind up and let go, because of the torque on the axle, the twist on the axle.*

LEON POND

Skidders enabled a lot of small loggers to become established in the business. A typical West Coast operation at this point consisted of a faller armed with a new McCulloch chain saw, a Lawrence 10-10 yarder and a skidder — a Kenworth, Clark or Timberjack — to haul from the landing to the dump at salt water. By the time logging activity had worked back more than a couple of kilometers from the water to where a truck was required, the logger had earned enough money to pay for it, as well as a road it could run on and a loader to stack the logs on board.

Conditions encountered by loggers on the West Coast also encouraged an evolution in logging-truck design. Elsewhere, where conditions were easier or public highways were in place, conventional trucks served the purpose. By 1950, trucks had almost replaced trains as distance haulers. That year, in the coastal region, only fifteen percent of the logs moved were handled exclusively by rail. Nine percent were moved by a combination of rail and truck, and the remaining 69 percent by trucks alone.

For the first decade or more after the war ended, the increased need for trucks was met by surplus military equipment, then widely available. But as the surplus equipment was used up, and the conversion from rail continued, coastal loggers were moving back farther into the steep valleys and higher up the slopes. They wanted bigger and more powerful trucks. This was a fairly specialized market, one that few conventional truck manufacturers were willing to enter. International, White and Mack were the major contenders from the mainline truck builders, with Kenworth, Hayes, Peterbilt and Pacific on the West Coast specializing in heavy-duty and off-highway trucks.

There was also an effort to develop a hybrid vehicle for hauling tree-length or whole trees directly from the stump, particularly over the kind of wet, rough ground found on the West Coast or in southern American swamps. Once again, the inventive Bob LeTourneau was involved. The LeTourneau Log Transporter was a cross between an off-highway — way off highway — truck and a gigantic forwarder. It had six custom-made, wide-faced, 1.8-meter-diameter rubber tires on independently-driven, electrically-powered wheels. When one wheel lost traction, all the power was transferred to the others. The machine went practically anywhere without the need of a road, bringing huge loads down 45 percent grades. A few of them were used through the 1960s, but they required a lot of maintenance and had a limited application. Bob Woods, a Vancouver Island contractor, used two Transporters in the late 1960s to pioneer Coastal whole-tree logging for Crown Zellerbach on some relatively flat ground at Sonora Island, north of Vancouver, British Columbia.
He hauled whole second-growth hemlock trees three kilometers over unroaded ground to the log dump, where they were limbed and bucked. The limbs and tops were laid over soft spots along the trail. That was the last time Transporters were used in British Columbia.

“

*I used a Skagit (cable yarder) that I bought on the West Coast. Then I told my boss that I wanted to have Taylor build me one. I said I can do just as good with a 60-foot spar that he'd build as with this Skagit. He said, well, go have him build you one. I put it on a tank, a used tank, and then powered it with a 313 Cat engine.*

*Well, the first one I built I tried to get him to put some hydraulic cylinders on it, to lay it down. And Taylor said, "We can't build a 60-foot spar. I can't build one heavy enough that would take that much pull."*

*"Yes it will," I said.*

*"N-o-o it won't." So he built a 50-foot, in about 1970, and I took it to Florida and pulled those swamps down there. It did so well I went back and told him I want one with a 60-foot spar.*

*"Okay. Yeah, we can build it." So they built me one with 60, and the more height you have, the further you can reach out. With just a little brake on it, you can pick those logs right up out of those stumps easy, you know.*

*And I told my boss, I said I want to build an 80-foot spar this time.*

*"No-o-o," he said. "you'all doing too good. You don't need a 80-foot spar." I think I bought six of them from Taylor and we had them all over. After he built them for me he beefed it up some and sold them everywhere. They called them a Taylor Prehaul Skidder. They were relatively cheap. Georgia Pacific's still got some of them running. You won't find anything that'll beat one of them pulling.*

Joe Watkins

During the late 1950s, a mechanization process began in the Soviet Union that was isolated from developments in North America and Europe. Prior to World War Two, the mechanization of Soviet logging was not a high priority because of the easy availability of labor — including a large number of female forest workers —and the existence of vast timber resources. At the beginning of the war, there were only seven chain saws, 126 tractors and eight logging railroads in the entire Soviet Union. At about this time, experiments were conducted with steam haulers pulling sleighs loaded with logs over iced roads. Delegations were sent from the Soviet Union to collect information on forest machines used in the US, Canada and the Nordic countries. But the war intervened, and this type of economic cooperation came to a halt.

After the war, the Soviet government was determined to become a leading exporter of forest products. The five-year plan adopted in 1946 doubled the number of forestry workers, and the number of forest machine operators increased by 500 percent. Many new lumber and pulp mills were established, creating an increased need for logs. Soviet rivers were of little use for timber-floating operations because they flowed in the wrong direction, toward the Arctic Ocean. Schemes to reverse the flow of the rivers were considered, but abandoned. To conserve scarce petroleum reserves, Soviet logging engineers developed portable electric generating plants to power electric chain saws, tractors, winches, loaders, cranes and narrow-gauge logging railways. They tested an electric farm tractor, the CTZ-NATI, which obtained power from high-voltage lines strung above the fields, along with a prototype tracked log skidder, the ETT-1.

The most successful development of this period in the Soviet Union was the KT-12 log skidding tractor, a natural-gas-powered, tracked skidder with a dozer blade, a winch and a tilting table at the rear upon which a turn of logs rested during skidding. It was a machine similar in concept to the various tracked pulpwood-hauling machines developed in Canada and Western Europe at about the same time to deal with similar conditions. The KT-12 was an extremely popular machine and thousands of them were used throughout Siberia as tree-length skidders. The wood-gas-powered version of the KT-12 had a gas generator which burned chopped wood as well as chips and leafy branches. Off-road vehicles made from ZIS trucks were also used as skidders.

During this period, there was little exchange of ideas with loggers and engineers outside the Soviet Union. Western buyers were not interested in the Soviet machines, an attitude fostered in part because Soviet productivity figures were only a fraction of those achieved in the West.

In the Nordic countries, progress was slow until the late 1950s. Throughout the decade there was a strong distrust of forest machines, particularly by horse loggers, who were looked upon as the nobility of forest workers. They were independent contractors and considered themselves part of the property-owning class; a horse was a big investment. The era was characterized by a clash between defenders of the horse and advocates of the machine. Horse owners considered it a victory when the machine builders failed; the machine people persevered.

“

*After the war we had the old Logger’s Dream. They revolutionized logging. They were built by Taylor Machine Works in Louisville, Mississippi. And old man Taylor, I asked him one time, “How in the world did you ever come up with the name Logger’s Dream?”*

*“Well,” he said,”we were over in Mississippi and we went out and a fellow was pulling some trees out of a deep hollow. He had a block in a tree and he had a pair of mules and he pulled a cable down and it pulled the cut-up logs up. I stopped and asked him, I said, Mister, would you like me to pull those logs up for you? And he said, yessir, I sure would. I said I’d pull those logs all out in a few minutes. And the old farmer looked at me and shook his head and said, Man, that’s a logger’s dream.”*

*And so that got to be the Logger’s Dream right then.*

Joe Watkins

“

*One of the things that I feel made as much difference as anything in the mechanization of the south were the foresters who were involved in the wood production system at that time. Many of them were junior officers in the military and they were World War Two vets. They came back with the attitude that systems can change. They didn't have the tradition of grand-daddy doing this and grand-daddy doing that. Or they saw that the old era was gone and they were in a new era. And what happened was they put the whole damn wood supply system into play and that brought improvement. When you look at the evolution in other countries — Sweden in particular — when they mechanized felling and they took the process of what a man with a bowsaw and an ax was doing in 1900. Now we have mechanized machines to do exactly the same process. We do the same stages but we do it mechanically. Because that was the way it was done. Here, we went through that period with the Busch combines trying to do the steps. You cut your tree down, you take the limbs off, you cut it up, you stack it and you carry it to the roadside and put it on a truck. Yeah, that was reproducing the steps that we did with a bowsaw, or an ax. But why do we do that?*

*So they started looking from the mills out, and said why do we want to cut it up in the woods when we can process the whole thing either at a railyard or at the mill? And this whole wood supply system of bringing it out in the short truck and loading it on a railcar and moving it in, that was not a variant. And so we changed the wood supply system to allow us to take advantage of the efficiencies in the woods. Went from 5 foot 3 inch wood to tree length.*

*The skidder helped, the knuckleboom loader helped. It was impossible with an old Idaho jammer or the Logger's Dream type loader we had then to handle tree lengths. You could do it, but it was difficult and dangerous. You took a lot of 5-ton trucks, put a fifth wheel on them, and ran a trailer behind. There was an evolution in trucks. Went from the used Pepsi Cola truck to the used gas-burner single or tandem-axle tractor with a trailer and then in the last 15 years to the new Macks, Peterbilts and Kenworths. So there's been an evolution in trucking, but that evolution came along with the move toward tree length material.*

BILL STUART

In Sweden, the education-minded labour-research institute (SDA) established its own forestry school, and 2,000 students enrolled to study the theory and practice of mechanical logging. The Finns quickly established good relations with their neighbor's research centers and an exchange of information was initiated. Jaakko Piironen, who was then a supervisor in one of the major Finnish forest owners' associations, realized the direction in which logging was moving and became the country's leading advocate of mechanized forest machinery. The number of agriculture and forestry workers was declining and would drop by more than 40,000 workers in Finland alone during the 1960s. The number of tractors used on farms was increasing rapidly, and the number of horses declining. The forest industry recognized the need for mechanization as forest-product exports climbed and the horse population fell.

The 1950s was a dramatic decade for loggers in many parts of the world, particularly North America. Within ten years many of the tools they used to harvest timber had become historical curiosities, others had been transformed. The new machines appearing in 1960 were far from perfect — in fact, many of them were merely crude prototypes, soon to disappear. But a new direction had been set and new ideas were being picked up and applied in other countries. For most of the world, it was clear that in the not-too-distant future logs would not be knocked down and dragged out of the forest in a series of disconnected operations. Instead, they would be extracted more carefully, with smaller, more sophisticated equipment employed in new integrated systems. During the following decade, these systems, and the machines which supported them, would begin to appear in the woods.

Horse logging crews thinning on a Weyerhaeuser tree farm in Washington in the early 1950s. The horses are hauling on a loading cable which runs through a block hung from an A-frame.

A LeTourneau Tournarch, a diesel-electric skidder, with articulated steering, skidding a turn of Douglas fir logs on the British Columbia coast in 1957.

The Skyhook, built by Phil Grabinski, was used in several locations throughout the North American northwest in the 1950s. It was suspended from parallel skylines and the operator lowered a skidding line to haul in logs.

A Madill steel spar has skidded logs out of a deep valley on the British Columbia coast. They are being loaded on a Kenworth truck by a Caterpillar wheeled loader.

A Big Stick loader loading pulpwood in the southern US in the early 1960s. This device removed much of the labor from pulpwood logging.

Loading Douglas fir logs onto a truck with a hayrack boom in Oregon, 1947.

An early model Prentice GOBC Loader used to feed logs into a Prentice Saw-Buck, one of the first production-model slashers made in the US.

A truck-mounted Prentice loader from the 1960s. This hydraulic loader was developed by Leo Heikkinen in Wisconsin, and helped revolutionize log loading.

The Loggermobile skidder, built by Wagner Tractor in the western US, was the first rubber-tired skidder with frame steering.

A Timberland Machines mobile cable yarder built in the 1950s for logging in eastern Canadian pulp stands.

A tracked version of the Logger's Dream for logging in southern US swamps.

This conventional model of the Logger's Dream was the most popular cable yarding and loading machine in the US South during the 1950s and 1960s. It was built by Taylor Machine Works in Mississippi.

The Timber Tosser loader was operated from a cable attached to the truck's driveshaft. It lifted logs from the ground and dropped them onto the truck's deck.

This original model Garrett Tree Farmer, introduced in 1958, was the first small, frame-steered skidder built that was designed to replace a horse for skidding logs from the stump to roadside.

A conventionally steered Timberskidder is used to haul a bundle of pulpwood. Timberland Machines' first skidder was introduced in 1956.

A Timberskidder skidding tree-length logs to a dump in Ontario in the late 1950s.

One of the earliest skidders, the Blue Ox was first built in 1952, and sold well in North America for the next ten years. It had front-wheel steering.

The Timberjack skidder, first produced in 1961, came with a small gas or diesel engine, and was an immediate success. It launched a revolution in logging machines.

Although designed as a log skidder, this Tote-A-Load skidder, built by Taylor Machine Works in Mississippi, could also be used to tow a Taylor log trailer.

The VSA forest tractor developed by Lars Bruun. It can be considered the forefather of Scandinavian forwarders. The VSA had frame steering, a hydraulic Hiab "Forest Elephant" loader and all-wheel drive. The back wheels were equipped with hydraulic motors. Carrying capacity 9.5 tons.

The first Timberjack skidder imported to Finland, purchased by Alf Swanson. The Swedish Domän company placed an order for two hundred of the same machines.

The cab of Alf Swanson's Timberjack skidder was rather chilly and bare compared to today's forest machines. In winter, a sheepskin coat provided heat, and in summer a string vest provided air-conditioning.

The late 1950s saw the onset of winch skidding in Finland and Sweden. The Swedes gave it up earlier than the Finns, who continued winch skidding into the late 1960s. Winches were installed on farm tractors and skidders. Winches attached to skids were transported from one work site to another by horse, by tractor or by the winch's own cable.

In the 1960s Valmet began producing skidders modeled on those built in North America. In Scandinavia, this development touched off a heated debate on full-tree versus cut-to-length logging.

Though at the beginning of 1960s many debarking machines were still working in Scandinavian forests, debarking was gradually moved to mill yards. Here a Sunds debarker.

Debarking machines increased in size, speed and performance. Initially, the spikes on feed rollers damaged the logs and were eventually replaced with chisel-shaped teeth.

At the beginning of the 1960s huge debarking and delimbing machines appeared in Swedish forests. This Sunds processing system handled the output of a dozen skidders, and required the use of a wheeled feeding machine and a loader.

# Searching for the Right Combinations

At the same time that rubber-tired machines were developed to skid whole-tree logs to roadside, another line of equipment evolved designed to process logs at the stump. The first of this new generation of logging machines appeared in December, 1957, when the earliest prototype of the Vit feller-skidder was finished. Rudy Vit was a young engineering student in Czechoslovakia during the late 1930s, and when the war began he joined the resistance movement. At the end of the war he found himself in Quebec, where, by the mid-1950s, he was engaged in a project with Bombardier and the North Shore Paper Company in Baie Comeau to design a mechanized timber harvesting machine.

The concept had grown out of a project launched at about the same time by an associated firm, the Ontario Paper Company, at Heron Bay, Ontario. At this operation a Caterpillar tractor equipped with an A-frame used a high-lead cable system to skid chain saw-felled trees to roadside where they were piled in bunches. One of the first Michigan PL-75 skidders took the bunches to a processor which delimbed them with a chain flail, cut them to length and loaded them on pallets. The bottlenecks in this system were at the falling and skidding phases. Vit's new machine addressed both these problems by making creative use of two recent innovations: chain saws and hydraulics. His falling component consisted of a compact unit mounted on a Bombardier HDW. An articulated falling head equipped with two chain saws was mounted on the front of the Bombardier. One saw made an undercut; the other severed the tree, which was held by the hydraulically-powered head. The tree was lowered over the machine and laid on a rack at its rear, which held several trees. When a full turn was accumulated, it was skidded to roadside. Eleven of Vit's Bombardier-mounted units were built over the next nine years, demonstrating the potential for a machine that could fall, accumulate and skid a turn of whole trees to roadside.

Because it was a full-tree harvester, Vit's machine was at least a decade ahead of its time. During the next ten years, almost all the effort devoted to the development of harvesting machines focused on shortwood harvesters that were able not only to fall and delimb, but also to buck the trees into short lengths to meet the demands of the established pulpwood handling process. To deal with the bucking phase, Vit designed a companion machine to process the trees skidded to the road. The Bombardier Processing Unit delimbed trees by winching them through a chain flail, after which they dropped onto a conveyer, were bucked to length and loaded onto a pallet. The processor incorporated several new concepts. The chain flail delimber, which had been developed a year earlier at Heron Bay, was new, though similar to one devised by Tomahawk Timber in Michigan at about the same time. Also unique to the Vit processor was an electronic measuring device which read the tree's diameter as it was being processed, measured the length of the logs bucked and computed their volume.

A year after Vit's prototype was finished, John Pope built another harvester at North Western Pulp & Power's Hinton, Alberta, operations. It was mounted on a crawler tractor and utilized a falling head similar to Vit's. The hydraulic-powered head gripped the tree while a chain saw cut it off. Pope's harvester then rotated the tree to a horizontal position and laid it into a delimbing head, which sheared off the limbs and pushed the tree back through the falling head for bucking. Logs were left in a pile on the ground at the rate of about three cubic meters an hour.

Vit's and Pope's machines, while revolutionary in concept, both suffered from the same drawback — they had to drive right up to a tree to fall it. Maneuvering for position took more time than any other phase in the cycle and severely limited production. In 1959, Jack Boyd of Timberland Machines offered the CPPA Woodlands Section a solution to the problem inherent in the Vit and Pope machines. The Timberbuncher, as Boyd called it, incorporated a turntable mounted on a tracked undercarriage, and utilized a knuckleboom with a grapple and saw to fall and pile trees in bunches. Although this was the first proposal to add the new type of boom to a harvesting machine, the feature which attracted the most interest was the falling device. It used a hydraulic cylinder to force wire rope through the tree. To everyone's surprise, this caused far less shattering of the cut end than the tree shears then available. Unfortunately, none of the forest companies were willing to finance development of Boyd's harvester, even though the knuckleboom concept eventually prevailed. Like Vit's, Boyd's machine arrived a decade too soon, in part because the right machines were not yet available to skid or forward the bunched logs or whole trees to roadside.

Another harvester, made at the behest of the Ontario-based Marathon Paper Mills, addressed the weakness in the Vit and Pope machines. The fabricator of this machine was Bob Larson, maker of the Hiabob in Ely, Minnesota. When he sold out to the Beloit Corporation in 1963, his machine was called the Beloit Harvester. Mounted on a crawler, Larson's harvester employed a distinctive new concept, a knuckleboom which could reach out from the tractor, delimb the tree with a set of knives, top it and fall it with a shear. The severed tree was then piled to one side. There was great interest in this harvester, and over the next ten years forty of them were built. Because the manufacturer increased its size to accommodate bigger timber, however, the Beloit harvester became so heavy and cumbersome its mobility was decreased beyond the point of usefulness. A later version of this harvester, developed by Beloit with Domtar, severed the tree first, delimbed and topped it in the vertical position, and stacked it upright in a rack at the rear of the machine. When a full rack of about forty trees accumulated, they were laid in a pile on the ground. Only one of these models was built, and shortly after it was destroyed in a fire Beloit abandoned the logging-equipment manufacturing business altogether.

Meanwhile, in Louisiana, Tom Busch completed development of his harvesting machine, producing a prototype in 1959. Designed to work in drier southern pine forests, the Busch Combine was mounted on a purpose-built, articulated, rubber-tired tractor. It drove up to a tree and gripped it while a shear cut it off. The tree was then laid in a carriage where it was delimbed and bucked to length. The severed logs fell into a cradle, where they were tied into one-cord bundles which were ejected for

"

*Factors outside the woods were what really led to the changes in mechanization. Labor cost, the interstate system had a helluva impact on the South, economically. We had a migration of labor out of the South, we had a migration of labor from the rural areas to the urban areas. We didn't need as much labor in the rural areas because of agricultural technology. So what we had to do was find ways to continue to produce the amount of wood with the labor pool we had available. And had to find ways to develop systems that required a modest capital requirement to match that labor. One of the ways you do that is to strip the functions to their simplest component. You want to cut a tree down mechanically? Don't hang everything in the world around it, get a shear and cut the damn thing down.*

*Share your base machines with the widest range of industrial or agricultural components that you can. We went to the Jimmy 353 engine, the Rockwell 75 axles, Eaton transmissions. You can get spare parts at most any truck stop or farm store. Everybody knew how to fix them. And because the design and development cost was spread across such a large base, the extra margin that went into the machines was relatively small. You go out and invent the Green Valleywopple, which has parts from all over the world, and you're only going to build three of them. The R&D costs have to be written off against three machines. But something like a skidder or a shear that's mounted on a small crawler or on a Bobcat — hell, every chicken farmer in the country had a Bobcat to move chicken shit. Why not hang a shear on it and cut down trees with it? Those systems matched our labor requirements and they matched our capital.*

BILL STUART

pickup by a forwarder. The Combine's two-man crew produced sixteen cords a day, compared to four cords for a top-flight two-man crew using chain saws and hand piling. The Combine performed well with trees from twenty to thirty centimeters in diameter, but shortly after it was developed a new type of sawmill technology, the Chip'N Saw, directed this size timber to lumber production. This meant most pulpwood was under twenty centimeters in diameter, too small for the Combine to harvest efficiently.

To retrieve the bundles, Busch also designed a forwarder called a Busch Loader. It was another articulated, rubber-tired tractor with a large hydraulic grapple that simply hoisted a bundle of pulpwood off the ground and carried it to the road. In later models the two machines were combined into one.

Even with its limitations, Busch's harvester-forwarder combination had several advantages over its competitors. It was fast and maneuverable, travelled on public roads, cost less to produce than a tracked equivalent and was cheaper to maintain. Its ability to oscillate as well as articulate made it highly mobile, and its delimbing configuration was much copied by other designers. Its shears were the most efficient on the market, even though the damage they did to frozen wood prevented much use of the Combine in northern forests. The machine caused a sensation among loggers, and spurred development of several other harvesters. Busch later produced a thinning harvester mounted on a Ford farm tractor converted to a three-wheeler, the TH100. It evolved into the TH210, which combined the best features of the Combine and the TH100, but only six units were built.

In 1962, Koehring-Waterous decided to build a harvester as its point of entry into the logging equipment business. The company's roots reached back to 1848 when Charles Waterous, an inventive engineer with a genius for organizing people and ideas, went to work for a small foundry in Brantford, Ontario. He established what later became one of Canada's largest machine manufacturing companies, with branch plants in the US. In 1953, the company was acquired by Koehring, a US firm with similar manufacturing interests. Ten years later, the company obtained rights to the Dowty rubber-tired forwarder and began work on a harvester using the Dowty chassis. As the design for the rest of the harvester's components proceeded, it turned out the chassis was too small and a larger, articulated version had to be substituted. The initial concept of the Koehring harvesting unit included a felling shear on a boom which reached out and severed a tree, holding it upright for processing. The tree dropped through a processing head which delimbed and bucked off logs which were stored in a rack and dumped out in one-cord bundles. A refined version of the harvester, the IIIB, held the tree at the butt and moved the delimber up the stem, which then gripped the top portion and bucked off logs from the lower end. The logs were accumulated and delivered to roadside.

The Koehring shortwood harvester was widely regarded as the most successful machine of its kind; by 1970, thirty-five units had been sold, most of them to Canadian International Paper, which was engaged in an extensive mechanization program. By 1977, when the last unit was

“

*In the late 1960s, through Timberjack, we were trying to get a harvester that could go into the swamps. The idea was being able to stay on top of the ground, float, and not destroy the ground and be able to move in low-ground areas. By this time the whole industry was using skidders. The horse had been phased out by then. We wanted a harvester to cut the tree down, delimb it right at the stump and put it into a bunch on the back of the harvester and then move along to the next tree, so that eventually we would have a cord, 15 or 20 trees, in a bunch on the back. We'd put a choker on it and drop it. Then a skidder would come and hook onto that choker and pick it up. This was going to be our mechanized way of doing things.*

*We were hoping to get our man-day productivity up from about five cords to about 10 or 12 cords. So instead of having a three-man crew, where you had a man felling the tree, a skidder skidding it and another man delimbing it, we had a two-man crew producing twice as much wood. The concept was great. They just didn't have the right materials, the alloys, at that time.*

Garnet Bell

produced, it had evolved into an incredibly sophisticated — and expensive — machine. Many observers believed this was its downfall. It incorporated so many improvements that it priced itself out of the market. However, the harvester firmly established Koehring as a major contender in the logging equipment business, and led the way to the company's highly successful feller buncher a few years later.

Throughout the 1960s several other prototype harvesting machines appeared, some of which went into production. In 1966, Vit built a new feller skidder, this one mounted on a heavier, wheeled chassis produced by Sicard, the Montreal subsidiary of a Seattle-area manufacturer, Pacific Car and Foundry. Of the forty-one units built, seven were shipped to France. This machine suffered from the same shortcoming as Vit's first model — it had to go right up to the tree — and production ended when Sicard left the logging equipment business.

An Australian-built machine, the Windsor Tree Harvester, built by R.L. Windsor and Sons in Queensland about 1968, was designed to thin pine plantations. It was mounted on rubber wheels and felled with a forward-mounted shear. The tree was then hoisted above the cab, limbed with knives and topped with a second shear, after which the logs were dropped in piles. It never developed into a production model, but during the 1970s, in cooperation with Timberjack, it evolved into the Timberjack RW-30, which also bundled the tree-length logs and dropped them on the ground for retrieval by a grapple skidder. The main problem with this machine, like many other harvesters, remained the need to drive right up to each tree felled. Without the availability of a hydrostatic transmission to provide precise ground speed control, a later development, this was a difficult and time-consuming task. Almost forty RW-30s were built before production ceased.

In 1966, Timberjack and Anglo-Canadian Pulp and Paper began development of a prototype shortwood harvester mounted on tracks which felled, delimbed, bucked and bundled pulp logs. It was field tested and a second prototype finished in 1970, after which the project was shelved. During the same period, Caterpillar made one of its early forays into logging equipment production, using a Model 950 front-end loader to carry a falling-delimbing-bucking mechanism. Lacking a boom, and mounted on a machine ill suited to off-road conditions, it was not particularly mobile. An uncomfortable machine to operate, it was abandoned in the mid-1970s.

One of the more unique machines of this type, which appeared in the southern US in a single prototype, was known as the Jesus Machine. It was developed by a team of forest engineers who wanted to set up a business to raise money for their church, but this was not the source of the machine's name. The harvester consisted of a scissor-type felling shear with some sort of hydraulic accumulator that enabled it to slice through a tree in a quick snip. The tree was laid horizontally onto the top of the machine and a set of exceptionally powerful rollers ran it through a delimber, while the shear snipped off logs of the desired length. The entire apparatus, weighing about 3,600 kilograms, was mounted on a Bobcat, which labored through the woods under its heavy load, constantly getting bogged down in soft spots.

On one occasion, Ken Patterson, a man of advanced years who developed a deep and abiding interest in logging equipment during his years at International Paper, went into the woods to inspect this machine, about which he had heard a great deal. Patterson was an elegant gentleman, known to visit logging operations immaculately dressed in a three-piece suit and polished oxfords — as he did on this occasion. On arrival at the site, he learned the harvester was not working properly; it could only fell and limb the tree, but not buck logs to length. That was fine with Patterson. The operator advised him to stand well back, and moved into position next to a 30-centimeter pine tree, about twelve meters tall. In spite of warnings, Patterson kept edging closer to get a better look. The harvester snipped off the tree, laid it horizontal, where it was automatically fired through the delimber by the machine's powerful rollers. With no bucking phase in operation, the log shot out of the machine like a gigantic arrow launched from a long bow, right past Patterson's head. "Jaysus!" he yelled, leaping backward a good two meters in one bound, as the limbless tree rattled off through the standing timber. That, apparently, is the origin of the harvester's name.

> *Early in the winter of 1960, we had three snow storms in Virginia, and the mill was running out of wood. We had to somehow log the wood into the mill faster. The way we were doing it until then was to fell the tree, drive these old bobtail trucks to the stump where guys cut them to length by hand. Then they lifted a 12-foot log onto their shoulders and threw it on the truck. Well, that was pretty damn slow productivity when you consider the mill was running out of wood.*
>
> *We couldn't get the trucks into the woods because the ground was muddy, and so I got an idea — I'm probably not the first one to get the idea but nobody in my region had thought of it. I hired every crawler tractor I could rent in a 50-mile radius and brought them down to the job. I put chains on the back of the machines and we dragged the trees to the roadside. Absolutely dragged the trees out and instead of the trucks going in the woods, the trucks waited for us to buck out by hand at roadside.*
>
> *Later that year they told me that on Monday morning we'd have one of those new log skidders coming on the job. No more bridges; we could walk right through the streams. And I said I'd have to see that. I drove out to the job that morning and there was one of the early Franklin Loggers, I think it was model 100. It had no blade on the front and it was very crude, but it did what they said. It walked through the creeks and streams without having to build bridges. And it had higher productivity. It skidded faster.*
>
> Mal Crawford

Several other shortwood harvesters were developed at this time which only made it to the prototype stage, or only had a few production models built. Bob Larson, whose eccentric mechanical abilities were employed on a variety of projects, built a rubber-tired shortwood harvester in cooperation with Great Lakes Forest Industries. It had a feller-processor head mounted on a knuckleboom which sheared the tree and laid it in a processor that used knives to delimb it, then sheared off logs at set lengths. The bolts were loaded into a cradle for transport to roadside. On an improved model, the cradle consisted of a hydraulic grapple and boom which unloaded a bundle of bolts in one movement. Although Larson's machine outproduced any others of its kind, only six were built because the pulpwood industry went into one of its periodic slumps and financial support for further development ended.

The Woodsmobile harvester, built in 1966, also used a knuckleboom with a feller-processor head mounted on a skidder chassis, but it piled its bolts on the ground. It was built by Northern Engineering and Supply Company of Fort William, Ontario, in a limited run of two pre-production units. At almost the same time, Nestor Siiro, an Angora, Minnesota, machine designer, built a prototype harvester with a similar processing head mounted on a Prentice knuckleboom which he set on an International crawler tractor. A significant feature of Siiro's processing head was a hydraulic chain saw, used for the first time in this application. In 1967, he sold his rights to Omark, which continued development and turned out four production units two years later. These machines were two of the earliest versions of the single-grip harvester which would be perfected in Sweden in the following decade.

Perhaps the last serious attempt to develop a harvester during this phase was undertaken in 1967 by Timberjack, in cooperation with the Spruce Falls Power and Paper Company in Ontario. This machine utilized an articulated skidder frame, with a felling, delimbing and topping mast mounted above the operator's cab. A grapple bunk — one of the earliest clam bunks — was located at the rear to hold the processed, tree-length logs, which were dropped in choker-wrapped bundles. Five production models were built and delivered in 1971, before the harvester was terminated — probably due to its high cost in comparison to other machines coming onto the market.

At this point a machine appeared which eclipsed the Koehring and Spruce Falls harvesters, and pretty much brought the development of full-processing harvesters to an end. The Drott feller buncher introduced a new harvesting concept that depended on the evolution of another machine, a new kind of skidder, for its rapid rise in popularity. This feller buncher was developed by Erv Drott, a long-time builder of logging equipment who sold his whole operation to the J.I. Case Machine Company of Wausau, Wisconsin, in 1968, the year his new feller buncher appeared. It was simple in concept. Built on an excavator base, a knuckleboom-mounted felling head, with a scissor-type shear, cut the trees and laid them in bunches. It could cut a fifteen-meter-wide swath at the rate of about one hundred trees an hour. For this ability alone, it threatened the popularity of hand-held power saws everywhere in North America except the West Coast. And with the development of a grapple attachment for skidders by Esco a couple of years earlier, eliminating chokers and enabling easy skidding of bunched full trees, the debate over where to process trees was more or less over. For the next twenty years, until the advent of Scandinavian cut-to-length technology, roadside processing systems dominated North American logging.

The evolution of Scandinavian forest machinery took an enormous leap forward in the early 1960s after more than forty Swedish forest professionals attended the World Forest Congress in Seattle. There, for the first time, most of them were able to see the various machines which had been developed in North America during the previous decade. The machine that attracted the most attention was Dwight Garrett's skidder, the Tree Farmer, which at that stage was hand-crafted in a shop near Seattle. The skidder so impressed one of the visitors, sawmill owner Sten Helmer, that he obtained the European-wide manufacturing and dealership rights to the machine for his company, Norsälven Sågverk AB.

At this time Swedish loggers considered the Bamse to represent the ultimate in forest tractors. Since it was first introduced it had been given a new red color, along with a more balanced weight distribution, stronger control equipment, hydraulics and better off-road capabilities. But, with the arrival of the Tree Farmer,

“

*We fiddled around and came up with what we called the three-man crew. A Timberjack 200 skidder — a little bitty thing — a middle-sized knuckleboom loader on a used truck, a decent log truck. I mounted a shear on the front of the Timberjack and on the back of it a grapple. Timberjack told me that by doing that I had made myself into the manufacturer and none of their warranties were any good. And I said, “Your warranties ain't worth a plug anyway. I'd just as soon be the manufacturer. Maybe I know what I'm doing and you guys don't!” We put together a three-man crew with those things. And right away, going to long wood, going to these things, the production of a three-man crew got to be 5,000 cords a year. We expected 5,000 out of them and we got it. You could buy a shear and you could buy a grapple. Timberjack then, I think after looking at what we did, begin to come up with shears and grapples. But the first one, we had a shear and a grapple that we mounted on the damn machine.*

*The skidder operator was the boss and he cut down the tree. A guy on the ground with a saw came along and limbed and topped. And when he had got a pretty good bit ahead with his cutting down, then he went back and skidded to a deck while the sawhand picked up the limbing and topping. The truck driver came and went and loaded his own truck. We had three people and they got 5,000 cords a year bang, bang, bang. Well, at that point, it worked. We did it just like you do it in the war. We froze the design, said fuck it, this is the way we're going to log, and we started putting on logging crews just like that. One year we put on 20 logging crews.*

*We made some horrible mistakes and we didn't have any maintenance facilities. But every dime I could get my hands on was put in something that would move a tree from the stump to the mill. At the same time we did that, in all of these little outside wood yards we put a Curry Cost Cutter — a saw where you drop the wood in a rack and the saw cuts up a whole rackful at one time. It is a miserable piece of equipment but it does one marvelous thing. It puts all the convenience out in the woods where it belongs. And it puts all the inconvenience in the wood yard, where it belongs.*

*I consider that a logger is like a rifle, anything he needs, bean or bullet, somebody needs to bring to him and lie it out next to him and all he's gotta do to get it is to look back and pick it up. He ain't supposed to have to go back for nothing. He ain't supposed to have to think. He ain't supposed to have to write letters. He ain't supposed to have to attend goddamn management intensification meetings. He's supposed to log.*

Tom Kelly

the Bamse's fate was sealed in the Nordic countries. Helmer's machine arrived in Sweden in 1961 and forester Robert Lundahl began testing it in Helmer's forests. The tests indicated the four-wheel-drive skidder had very good off-road capabilities. But ground skidding was not a method favored in Scandinavian forests. Consequently, engineer Arne Anderson converted the machine to a forwarder by removing the cable winch and installing a trailer and a cable loading boom. The loader was later replaced with a hydraulic Hiab Elephant, which had become Sweden's number one loader shortly after it appeared.

More North American skidders were imported that same year. Forester Folke von Heideken bought two Tree Farmers and Domän bought three Timberjacks, which were tested at Jämland. Opinions were divided over the new skidders. In addition to disliking the log-skidding method, critics argued that the chokers attached to the logs for skidding were expensive and hard to find in the snow and brushwood. (Tree planters had to be paid one krona for each choker found.) Those who liked the machines said they were easy to drive, reliable and had good balance and off-road capabilities. Because of their simple construction and few moving parts, they required fewer repairs than Swedish-built machines. North American tractors had the power and off-road capabilities that were desired at this time in the Nordic countries; they dictated the direction of the next wave of development, which started in 1961.

That year, Lars Bruun produced a newly-designed forwarder inspired by the Harrison Pulpwood Harvester christened the Brunett. The machine had several features new to Scandinavian equipment: four big rubber tires, frame steering and hydraulic-powered rear wheels. It also incorporated a Hiab Forest Elephant, consisting of a five-meter-long hydraulic loading boom and grapple. It was built on a BM 350 Boxer farm tractor with the front wheels removed, and carried nine-and-one-half tons. As was often the case when Bruun was involved, it sparked a heated debate over the issue of wheels replacing tracks.

Big forest companies quickly took an interest in the Brunett, and the orders started rolling in. In 1964, Ford tractors were used in its fabrication and, as a sign of cooperation, the Brunetts were painted Ford blue. Hundreds were sold during the next few years. The machine was exceptionally durable; at a time when twenty-five percent of a typical forest tractor's work time was spent on maintenance and repair, the Brunett got away with ten percent. It boosted faith in mechanization and silenced opponents, particularly the horsemen, who had laughed at the mechanized loggers, suggesting that their real work was not harvesting timber but repairing machines in the woods.

Stories of new logging machines across the Atlantic spread throughout Sweden, and many tours were organized to analyze the US and Canadian systems. Among those studied were the Bombardier HWD, the Michigan skidder, the Dowty forwarder and a giant grapple installed on an Allis-Chalmers tractor. The Busch Combine caused a big sensation, and the Swedish Mechanization Foundation quickly acquired two similar machines, a Vit Feller Skidder and a Beloit Harvester. A demonstration tour of Sweden was organized for both machines, and while neither sparked great enthusiasm they heightened the interest in mechanization. The pros and cons of the two North American systems —processing logs at the stump, which the Swedes called the "cut-to-length" method, and full-tree skidding to roadside — were systematically examined. Most importantly, ideas for a domestic felling machine started taking shape on many drafting boards.

During the same period, the Finns followed the same course as the Swedes, familiarizing themselves with Canadian and American systems. The new knowledge learned was put to the test in 1961 when a major windstorm felled big areas spruce forest in central Finland. American cable harvesting systems were tested in the tangle of fallen spruce trees, but proved too expensive and unsuitable. Finland started experiencing its own shortage of horses, which touched off a brief interest in cable yarding systems. Self-powered skidding donkeys were moved with tractors along skid roads. They were soon followed by line machines, the Finnish name for a farm tractor with winching equipment and skidding tongs. The line machines yarded logs from the stump to the skidding road up to half a kilometer away, their maximum skidding distance. Logs were loaded on trucks equipped with hydraulic loaders. There were hundreds of line machines in use until the mid-1970s.

“

*I was plant superintendent at the time the Timberjack skidder was developed. We were looking for something smaller (than the Timberskidder) to replace the horse. Basically, everything in the woods was done by a horse. In cable logging they used horses to get the logs over to the main line so they could hook on to them and skid them down to the road. So they wanted something to replace the horse. That's when they got into the articulated frame. I believe there was a unit that was made in the west, the Garrett Tree Farmer. I think they were on the market before Timberland started the Timberjack type of vehicle. I think there was some brought east to some of the logging shows and I think that's how the idea evolved from there. "Somebody else can build one, so can we!"*

*The Timberskidder, the big one, had a big Braden winch on the back end of it that used five-eighths cable for the haul line. That was used for tree-length skidding. They'd winch them in and scoot them down the road when they had to go a couple of miles or something like that. The Timberjack was not designed to replace the Timberskidder, but to replace the horse. That's why we started off with a small Ford gas engine, or the same size in diesel. The Timberjack was a lot smaller than the Timberskidder. It was a lot smaller than the units that they're building today.*

*The Timberjack and the Tree Farmer were both about the same size originally. They started off with a small unit to replace the horse and then they started getting bigger and bigger.*

Don Small

Their popularity declined when linemen began refusing to accept the low wages involved. A lineman's work included wading through the snow to the log, attaching the tongs and wading back behind the log to retrieve the line for another log, while the owner operated the winch in the comfort of the tractor.

The mechanization of reforestation in Finland also started in the early 1960s with the development of scarifying machines to mix the organic surface soils with deeper mineral soil. Workers could no longer be found for hand scarifying, so ideas were imported from Sweden. Under the direction of the Central Forest Association Tapio, the first Finnish scarifying machine was built, the hydraulic Tume-Tapio. It was mounted on farm tractors, four-wheel-drive Unimog trucks, Alftaband-equipped crawler tractors and wheeled loaders. In 1963, Finland got its first Timberjack skidders, which were also equipped with scarifying machines.

During the early part of the decade, horses and men were still available for logging in Finland, but production levels were on the rise, increasing at a rate of two million cubic meters per year. The power of the farm tractors available for transporting logs was inadequate, and the first Timberjack and Tree Farmer log skidders imported for testing did not make a favorable impression. On the other hand, the Swedish Bamses and Nalles, and Finland's own solutions seemed better suited to Finnish forests. The one place where excitement over the American machines ran high was at the Finnish state-owned tractor factory, Valmet. Particularly enthusiastic was Olavi Sipilä, an engineer at Valmet. He realized that the factory was in the best position to start building forest machines because it had the expertise and many of the component parts. Valmet unveiled its first four-wheel-drive forest tractor, the Valmet Terra, in 1962. A year later the company built a skidding tractor which never advanced beyond the prototype stage, but clearly it was interested in the sector.

In 1962, a thirty percent currency devaluation shook Finland, weakening those sectors of industry that relied on machines or raw materials from abroad. Buying foreign forest machines was out of the question, so Finland had to rely on its own manufacturers for awhile. The favorite forest machine of the day was any type of four-wheel drive farm tractor modified with half or full-tracks, an added winch, a hydraulic loader and a trailer. The Hankkija Co-op built Robur and MF-Kärppä forest tractors out of Massey-Fergusons it had imported earlier. As well, modified Ford-Countrys were popular in northern Finland.

Meanwhile, in Sweden, the people at Volvo and Ösa proudly introduced the new BM Volvo SM-360 forwarder, nicknamed "Big Bear." It overcame the weaknesses inherent in tractors adapted to forest work. The hard work took a toll on the tractors' axles, and their wheels could not take the stress caused by the tracks installed over them. A new type of heavy-duty axle was developed for Big Bear, along with frame steering, a better type of track, and a well-designed trailer. Big Bear was the first of the large-sized forest tractors, the prototype of contemporary Scandinavian forwarders. It was also an instant success, and when the premier of the Soviet Union, Nikita Khrushchev, visited Sweden he was so impressed by the machine he ordered one on the spot. Big Bear was easily modified, and became the all-purpose machine of Swedish forests. One of its adaptations was a skidder called Big Bear 661, which was equipped with skidding tongs. A smaller version of the Big Bear was made in 1966 for shorter hauls, and an even smaller version, Little Bear SM 460, sold extremely well.

With Ösa's help Volvo began modifying its farm tractors and earthmovers into forest machines. First of all it introduced the Timmer-Kalle and the Long Lisa forwarders to the market; and a short time later the Short Lisa skidder appeared. The Timmer-Kalle was probably the most significant of these machines, even though it was slow and clumsy. It had an air-tight, warm cab with a flat floor that was considered the height of comfort and progress.

The Nordic forest machine market picked up through the 1960s and, in addition to Ösa and Volvo, the business began to interest the big metal companies, such as Kockums, Stensele Mekaniska, Mo-Do Mekan, Sunds and MS-Robur. Stensele Mekaniska had been founded by the Sandberg brothers, who inherited their father's blacksmith shop in the 1950s and built it into an engineering workshop specializing in forest machines suited to conditions in northern Sweden. They sold the manufacturing rights to their first crawler tractor to Hagglund & Soner, and in 1965 brought their first rubber-tired forest tractor, the six-wheeled Drivax, onto the market.

New competition also was kindled among hydraulic lift manufacturers. Hiab's Elephant soon faced new rivals; like the Vindelbjörn, which was later renamed the Cranab. As the decade wore on, American machines made a successful appearance on the market. Kopparfors AB's logging sites, for example, used primarily Timberjacks. Kockums started manufacturing Tree Farmers under license at its Mekaniska Verkstad AB in Malmö; Domän quickly bought eighty of them. Ösa took up the North American challenge, building two skidder prototypes, called S-1 and S-2, but its reluctant partner, Volvo, did not believe log skidding had a future in the Nordic countries so the project was shelved.

By the middle of the 1960s Nordic forest machine manufacturers were on the verge of an enormous increase in the development and production of an entire range of logging equipment. The focus of this development tended toward what became known as the Scandinavian system — the processing of logs at the stump and their transportation to a main haul road by some means other than skidding. Across the Atlantic, North Americans had also focused on a new direction: falling and bunching whole trees, which were then skidded to roadside for processing. Over the next twenty years or more, the two systems evolved with some degree of independence, before merging to produce a global logging technology.

The Koehring Shortwood Harvester, shown working in New Brunswick, Canada, was first built in the 1970s and was perhaps the most successful machine of its kind.

The Busch Combine, developed in Lousiana in 1959, was designed to fall and process pulpwood in southern US pine forests.

This model of Rudy Vit's first feller skidder, usually called the Vit Feller Buncher, was mounted on a Bombardier. First built in Quebec as a prototype in 1957, it was a harvesting machine well ahead of its time.

The Busch Loader was built to retrieve bundled pulpwood and carry it to roadside. Eventually, it and the Busch Combine were combined into a single machine.

Skidders and the full-tree method were still used in Scandinavia at the end of the 1960s, but the mainstream was already turning toward forwarders and the cut-to-length method.

The Vit Forwarder was built in Quebec in the late 1960s and early 1970s by Rudi Vit. It utilized a Lokomo drive unit to power a tree-length hauling machine, 78 of which were built.

The Lokkeri was Lokomo's first forwarder. Its size and price horrified farmers, who were used to the cost of farm tractors. The cab was in front, giving the operator good visibility.

Kockums started building forest machines at the beginning of the 1960s. The factory initially manufactured Tree Farmer skidders under license. The first Kockums model was the KL 860, pictured here.

Kockums' Skruven delimbing machine wasn't a success. Logs had to be removed from it quickly before it chewed them into thin stems.

A host of new manufacturers swarmed onto the Scandinavian forest machine market in the mid-1960s. The Sandberg brothers' Stensele Mekaniska Verkstad quickly conquered the market in northern Sweden with the Drivax.

Lars Bruun began designing forest machines for Kockums in the late 1960s, such as the Kockum Bruunett 836B shown here.

Ösa worked with Volvo throughout the 1960s developing new machines. One of the first was the "Little Bear" 460, which had a loader mounted on the cab roof. The rear wheels had hydraulic traction.

The Timmer-Kalle reflected Volvo's design philosophy, which was to utilize as many of the company's farm tractor and earthmover components as possible. The Timmer-Kalle was also one of the first forest machines in which operator comfort played a role in cab design.

The cab of the "Big Bear" 661 showed signs of comprehensive design. The cab was operator-friendly by standards of the day.

The “Big Bear” and “Little Bear” tractors could be converted easily to a variety of uses. Here, an “Big Bear” 661 is equipped with a felling head and skidding bunk.

Reforestation became increasingly common in the 1960s in Scandinavia, and engineers focused their attention on developing planting machines. It was a daunting task because the seedlings required gentle handling. Pictured is a Swedish SHS planting machine, built in the 1960s.

The Swedish company, Hede Mekaniska, entered the forest machine market in the late 1960s and produced the Hemek All-driv. The forward wheels of the rear axle were powered, and power was transferred to the rear wheels with a roller.

Valmet developed the 882 feller-skidder during the 1960s; it had a felling head and skidding tongs. While it was an efficient full-tree machine, it was forgotten as the cut-to-length method became more common.

The Beloit Harvester is demonstrated at an equipment show in North Carolina in 1964. Built in Minnesota by Bob Larson, it delimbed and topped a tree before falling it.

The Drott Feller Buncher, which first appeared in 1968, revolutionized mechanized timber falling. The falling boom could cut and pile full trees in bunches for skidding with a grapple skidder.

The Timberjack RW-30 harvester evolved from the Australian-designed Windsor Tree Harvester. Forty of these machines were built in the late 1960s and early 1970s.

The Canadian-built NESCO Woodsmobile 100 Harvester, mounted here on a Timberjack skidder, had a falling-bucking shear and a delimber mounted on a knuckleboom. It, like the US-built Siiro-Omark harvester introduced about the same time, was one of the first single-grip harvesters built.

Scandinavia's first felling heads, such as the Garp grapple and the STFI shown here, worked like a big pair of nail clippers. The hydraulic powered felling head had two blades.

The Timberjack Shortwood Harvester was developed in 1966 with Anglo-Canadian Pulp and Paper in Ontario. After two prototypes were built and tested, the design was abandoned.

Timberjack's Spruce Falls Harvester was, perhaps, the last North American harvester built. Five machines were built before its design was eclipsed by the Drott feller buncher.

The Massey Ferguson Treever, a pulpwood forwarder which appeared about 1970, was one of the farm-equipment manufacturer's rare ventures into the forest machine market.

Sweden's first processor, the VSA, was designed by Lars Bruun in 1966. It delimbed and bucked logs.

Kockums also made a quick entrance into the harvester market with this model designed by Lars Bruun.

Finland's first processor, the Pika 50, designed by Sakari Pinomäki and installed on a Valmet carrier. It has been said that Pinomäki's processor started the second wave in the mechanization of Finnish timber harvesting.

# Improving on the Details

The mechanization process underway in the North American logging industry during the 1960s and 1970s did not go smoothly. Often the motivation for a logger or a large forest company to mechanize woods operations arose out of a set of circumstances external to the industry altogether. Wages, for example, were increasing in other sectors of the economy. In order to obtain and keep good workers, owners or managers of logging and forest companies had to offer higher rates of pay. The only way this was possible was to give their workers better tools, help them develop the skills to log more efficiently and increase daily production.

Through the 1960s and much of the 1970s, it was not always an easy matter to choose the right machinery. It was a period of rapid mechanization of every phase of logging, and many of the machines were capable of performing more than one task. It was less difficult to choose the right machine than it was to choose the right combination of machines. And to complicate matters, an entire line of machinery — such as multifunction harvesters — could be made redundant almost overnight by the development of another type of equipment.

Around 1963, rubber-tired skidders began a rapid evolution which continued for the next fifteen years or more. In retrospect, the evolutionary advance was not limited to the skidder, but included a whole line of logging machinery utilizing rubber tires and advanced hydraulics. What made such rapid development possible was not just improvements in the design of the machines themselves, although that was a factor. As important were improvements to two components of the machines, more or less beyond the control of skidder manufacturers — tires and hydraulic systems.

In the late 1950s, when the first skidders began to appear, the only hoses and tires available were intended for use on agricultural, construction and other types of industrial equipment. In these applications, they were usually adequate, but when they were used on logging machines, in many cases by operators without extensive mechanical experience, they did not stand up. A hydraulic hose capable of years of service on a farm tractor could be torn from a skidder or a feller buncher by an errant limb or a misfallen tree during the machine's first day on the job. During the first decade or two that hydraulics were used in the woods, the real beneficiaries of this technology were the manufacturers of hydraulic fluid, since a virtual river of it leaked onto the forest floor. Part of the solution lay in the design of machines with well-protected hydraulic lines, but part also lay in the development of tougher hydraulic hoses and more advanced controls. Perhaps the largest single contributor to the hydraulic revolution was a simple part worth only a few cents: the rubber O-ring, which made possible the tight seals required for sophisticated hydraulics. What aided in obtaining these improvements was the maturation of the equipment manufacturing sector and its growing ability to work with component manufacturers in developing improved products.

Perhaps the best example of this process had to do with tires. When skidders were first developed, the only available tires were made for farm tractors, and they proved inadequate for a log-skidding machine. Being narrow, they cut deeply into soft ground. Their treads were poorly configured for use in the woods, and were easily torn off. Their thin sidewalls — perfectly adequate on a farm field — were easily punctured by sharp sticks and stones. Their shape where they met the wheel rim allowed limbs and other chunks of wood to wedge between the tire and rim, breaking the seal and deflating the tire.

Bob LeTourneau tried to solve this problem by ordering custom-made tires for his huge machines, which cost a lot of money and helped price them out of the market. Other manufacturers tried to persuade tire makers to build tougher tires. This was not an easy task. For the large tire companies, in 1960, the skidder tire market was minuscule, too small to warrant the research and development costs involved. And to complicate matters further, skidders were in a rapid state of development, so the specifications for the tires they needed were constantly changing.

Among the first tire companies to recognize the potential of the skidder market was Goodyear, largely through its Ontario operations, in cooperation with Timberjack. Morris "Doc" Cryder, Goodyear's technical superintendent, worked closely, first with Bob Symons at Timberjack, and then with other machine manufacturers and engineers, to develop a tire to meet loggers' needs. It was not a simple process. Persuading a company the size of Goodyear to build, say, forty-eight tires of a new design, requiring new molds and layup procedures, was hard enough. Part way through the process, to change the specs on these forty-eight tires, was something else again. What made it possible was the growing relationship between people at Timberjack and the senior people at Goodyear.

Gradually, the tires improved, and skidders became much more effective at their job, naturally increasing the demand for skidders. Before long, other tire companies — especially Firestone and United — recognized the market potential and began developing their own tires for logging machinery. As the machinery began moving onto new types of ground — swamps, for example — new tires were needed. United led the way in developing wider tires to increase flotation and lower ground pressure. Loggers required low pressure tires to reduce the damage to the forest floor or to the next crop of trees that was beginning to appear as the original stands were logged. By the time four to five thousand skidders a year were being purchased in North America, tire companies acknowledged this new line of machines with its special tire needs, and were willing and able to supply the market.

Through the 1960s and well into the 1970s, there was a dramatic increase in skidder production. Orders for forty or fifty new machines were common. Established builders, such as Timberjack, were hard pressed to meet the demand. Timberjack's skidder sales went from 44 machines in 1961 to 1,500 in 1966, to more than 2,000 in 1973. This kind of expansion attracted newcomers. Dozens of new builders fed the ravenous market as loggers across the continent, and in many other countries, flocked to the iron mongers to get one of the new machines. As well, loggers being the independent-minded and inventive kind of people they are, numerous shop-built skidders appeared in the woods. In 1961, for example, Bill Hoffman of Hawthorne, Florida, put together an articulated skid-

"

*That first skidder was built from used parts. I had a Carco winch that I took off a crawler tractor and a Ford engine. Actually, what I did was I bought a six-wheel drive army surplus truck and used the axle and the engine. Probably stuck a Fuller transmission in it and made a reversing box so you could go backwards same speed as forward. And used truck tires on it. But it was just too heavy. I think I was up about 19 or 20 thousand pounds. Today, you know, everything is above that.*

*When I started out, you had to go where the crawlers couldn't dream of going. In order to do that you had to keep your weight down. The second unit that I built was made from John Deere combine parts. Had a huge reduction in weight. It could walk on water, and you hardly made a footprint but you couldn't keep it together. It was made for a combine that traveled in the field all the time. You couldn't overload it. Between those I figured I needed to come in around 10 to 13 thousand pounds. The next one came in right about 12,000 pounds. Today most everything you build is from 20 to 35 thousand pounds.*

Roger Drake

der made entirely out of Chevrolet truck parts, which he used to skid tree-length pulpwood to roadside. At about the same time, the White Owl Parts Company in Kinston, North Carolina, put together a rigid-frame skidder, called the Short Dog, made of surplus parts, which it sold for $1,100. The company sold several dozen before its skidder disappeared from the business.

This was the fate of many skidder companies founded in the 1960s. Pettibone, in Duluth, marketed a line of Master Speed skidders beginning in 1960, including one of the first grapple skidders, then abandoned the field. The Frank Hough Company, of Libertyville, Illinois, which later became part of International Harvester, began producing the Paylogger skidder in 1963. Gulf Engineering in Truro, Nova Scotia, had a skidder on the market for a short time, as did Anthony Manufacturing in Springhill, Louisiana (the Whirlybird), Kelowna Machine Works in British Columbia (the Timber Toter), the Kenworth truck company in Portland, Oregon, and Napco Industries in Minneapolis, Minnesota (the Napco Crab). Emil Gafner, who operated an engine rebuilding shop and a farm tractor dealership in Escanaba, Michigan, after the war, and who in 1965 brought out the highly successful forwarder known as the Iron Mule, briefly produced a Mini Skidder beginning in 1968. Taylor Machine Works, builder of the Logger's Dream, produced a skidder, the Tote-A-Load, as did Athey, Massey-Ferguson and a score of others, all of which have long since disappeared from the market.

But many others survived. One was the Franklin Logger, designed and built by Roger Drake, of Franklin, Virginia. Drake, who operated a loggers' service shop under the misleading name Cavalier Auto Service, perceived the need for a lighter, less expensive skidder. In his free time, and with parts scrounged from a variety of machines, he cobbled together a couple of prototype skidders — light, mobile machines similar to the Garrett and Timberjack models of the same era. After getting sidetracked for a brief period as a Pettibone dealer, Drake decided to take a serious stab at producing his own skidder, the first of which, the Franklin Logger, appeared in 1962.

One of the things that helped propel Drake's company, Franklin Equipment, into the big leagues of the skidder business was the close attention he paid to loggers' needs. This was apparent in 1965 when he was approached by the local Union Camp woods manager who was looking for solutions to the problems of logging in the Great Dismal Swamp in southeastern Virginia. The result of that conversation was the Swamp Buggy, the first skidder designed to work in the southern wetlands. It had a smaller engine than usual, a 97-horsepower GM diesel, and was equipped with Goodyear's newly-developed, 86-centimeter-wide skidder tires. The idea was to keep the machine light enough on the ground so that it would ride on the root and vegetation mat lying under the swamp surface. The Swamp Buggy operated with water halfway up its radiator. Franklin went on to become one of the industry's leading equipment manufacturers.

Also during this period, three of the world's major equipment manufacturers entered the skidder business. John Deere, well established in farm equipment and crawler tractors, came out with its first articulated skidder in 1965, the Model 440. From this beginning it went on to become the largest producer of skidders in the world. That same year, Clark, which ear-

"

*In the 1970s, hydraulic hoses were the weakest link in processors. One of my customers found this out the hard way. The main pressure hose erupted and the hydraulic oil escaped from the tank. The closest road was a kilometer away, and the other machines were still at another logging site. The contractor walked several trips, hauling a new hose and more than 50 liters of hydraulic oil for the machine.*

*As if to appease the city dweller's envy of working at the bosom of Mother Nature, the contractor managed to run his machine for just fifteen minutes before the trunk of an alder pierced the radiator. Of course, all the coolant leaked out. When I arrived on the scene the man had already carried dozens of liters of coolant from the road to the machine, and he was hissing: "You don't happen to know where I could get new, shorter arms. My fingertips are worned, because my arms got longer doing that carrying."*

Kauko Papunen

lier had abandoned its skidder line, came back into the market with four models. And after several years of development, Caterpillar produced the 518 Wheel Skidder in 1971. It was an articulated machine, built around an experienced Caterpillar drive train, and qualified as the company's first serious attempt at capturing some of the logging equipment market.

There was even a North American revival of the tracked skidder. Earlier attempts to develop tracked forest machines had encountered serious difficulties. There have been two basic track designs. The first, utilized by most crawler tractors, has a flexible rail connected with pins and bushings, with plates or grousers bolted to the chain to engage the ground and support the machine. A sprocket at one end, inside the rail, propels the machine, while an idler keeps the track tight. These tracks were prone to stretching and tended to fall off when working among stumps or rocks. The second type of track, used on Soviet forest machines and tanks, consists only of plates or grousers held together with pins and bushings which in later versions were made of rubber for flexibility. A set of three or four wheels, independently suspended, fit inside the tracks and propel the machine. Bombardier produced a series of forest machines using these tracks, but none of them were a great success, largely due to track problems. In 1974, Food Machinery Corporation (FMC), a large diversified US equipment manufacturer, brought out a new, high-speed, low-ground-pressure tracked skidder with improved tracks that performed well on soft ground and steep slopes. When FMC stopped producing the machine a few years later, manufacturing was taken over by Kootenay Equipment in British Columbia. Twenty years later, Kootenay builds twenty-four tracked KMC line and grapple machines a year, half of which are sold in Canada and the rest abroad.

A skidder accessory that had a major impact on the development of the machine, and on logging systems generally, appeared on the market in 1966. The hydraulic grapple, manufactured by ESCO, an Oregon company, turned the skidder from a two-man machine — or one-man machine with a lot of inoperable time while the driver set chokers — into a much faster, one-man machine. Turned 180 degrees, as Timberjack did on its Spruce Falls Harvester, the grapple provided the machine with quite a different set of capabilities, which were not fully exploited until a couple of decades later with the widespread use of clambunks. Beloit, which briefly produced a skidder called the Logging Tractor, was the first to come out with a grapple skidder, in 1965. With the development of feller bunchers, grapple skidders quickly dominated the skidder market.

During this period, the established skidder manufacturers expanded their product lines by introducing new models to supply gaps in the market. Timberjack, for instance, began production in 1961 with a 200 series, added a 300 series in 1968, a 400 series in 1969, and a 500 series in 1972, adding progressively larger models to meet demand.

Throughout the 1960s, a unique organization in Quebec, Logging Research Associates, focused on the development of a complete logging system that led the North American industry further toward roadside processing of whole trees. Funded by the Abitibi and Canadian International paper companies, LRA's Doug Hamilton designed a tracked skidder that eventually evolved into a wheeled feller-skidder called the LogAll. It had a felling-loading head mounted on a knuckleboom which severed the tree and loaded it into a rear-mounted clam bunk. LRA also produced a compatible roadside processor, the Arbomatik, to delimb and buck. Through a licensing agreement with Warner and Swasey of Cleveland, Ohio, fifteen processors and even more skidders were built for the research group's sponsoring companies. In 1971, during an industry downturn, the operation folded. The significance of the LRA machines was that, for the first time, a logging system was available which did not require anyone to work on the ground. Falling, bucking and delimbing with chain saws were all eliminated, as was manual choker setting. It was probably the first fully mechanized logging system.

Having struggled to catch up with North Americans since the war, Scandinavians were now working on similar logging systems. By the mid-1960s, tractors had replaced horses in Swedish forests. The most advanced logging operations were well equipped with machines. Debarking and delimbing machines worked at landings, and modern forest tractors skidded or hauled full-length trees to them. With all this processing occurring at central locations in the woods, it often looked like factories had been built in the middle of the forest. But something was still missing.

The harvesting machines that visiting Swedes had discovered in North America had made an impression. Throughout Europe and the Soviet Union, cutting and delimbing still were done mainly with chain saws. Although safe equipment and work methods had been developed, chain saws took a heavy toll on their operators. The vibration was bad for blood circulation in the fingers, eventually causing the so-called White Fingers Syndrome. The noise and heavy labor caused spinal problems, wear on the joints, hearing loss, tennis elbow and sciatica. The time was ripe for a Nordic felling machine, and in 1962 the Garpenberg Forestry University began work on a mechanized felling head. A couple of years later, a prototype was ready. It worked like a pair of large hydraulic nail clippers and was called Garpnäven, the Garp Grapple. It was the first European-built harvesting head.

Sweden's first processor was introduced soon after by VSA engineer Lars Bruun. The Bruunett processor had a long boom, a loading grapple, delimbing blades and a circular saw. Now it was possible to process logs at the stump in a manner consistent with traditional cut-to-length practices. Kockums introduced a second Swedish processor, the Logma, in 1967. Importing a concept from Canada, it had a long boom that fed a log through the debarking knifes before finally cutting the top off. The Logma had two cabs, one with a set of controls for moving the machine and the other for operating the processor. Critics of the Logma pointed out that it was only a delimber but, along with the Bruunett, it was an evolutionary step toward a falling head. A short time later, Kockums built another delimbing machine, the Skruven (screw), which got its name from long rotating bayonet-like spirals. It operated like a pencil sharpener gone mad; delimbed logs had to be removed from it promptly or they would be reduced to thin saplings. The Skruven never made it past the prototype stage.

"

*The tire problem was penetration by the slash as you drove through the bush. You couldn't go very far without poking a hole in the tire. And when you are out logging you don't just phone down to the corner and get a service truck to come out. So that was a major problem.*

*Doc Cryder undertook to design a skidder tire but he had to use available equipment. Therefore he started with farm tractor tires because that was the kind of equipment that was in the bush anyway. And so he just changed the construction of the tire. Basically, he went to a heavier carcass with nylon in it. Then eventually switched the compounding over to tougher compounds. Farm tires have to have certain characteristics but this new tire had to have a tough carcass so that it would give a little bit instead of getting impact breaks from this slash. He had to put in tougher compounds, more like off-the-road tire compounds. So that was his approach. He went from a carcass that would have perhaps four plies, usually rayon at that time, to ten plies of nylon, which made it much stronger. Nylon was more flexible, so if you hit something it would tend to push it out of the way.*

*Then he went to the more tear-resistant and more penetration-resistant truck tire compounds. We added a shredded wire undertread and a shredded wire sidewall. That's where you took little half-inch, three-eighths to half-inch, pieces of wire, extruded wire like piano wire, except very fine wire, and it was all chopped up and mixed in with the compound and put on as a layer underneath the tread so that when something hit it, it would act like a shield.*

*The next step — because the slash would trap itself between the rim and the sidewall, it would get hung up in there and sometimes just push the tire off but most times damage the tire — we designed a flange cutter. We recut the molds so that when the sidewall fitted over the flange, you had rubber coming down and it was just a deflector. They used that on aircraft and this is what we did on the log skidders. That worked well.*

Bob Partington

Ösa continued development of the Garp Grapple, replacing the shear with a hydraulic chain saw, then mounting it on a hydraulic boom attached to a sturdy forest tractor. The new felling head was sold to several forest companies and independent contractors who mounted them on a variety of machines. In 1968, Ösa and Volvo introduced a new forest tractor, the SM 668. It was an attractive machine with a molded cab, a clam bunk and hydraulic booms. It had six wheels and a tandem axle, and the heated cab was equipped with big windows. Operators said running it was almost like working in an office. The original engine was a Volvo BM-1113, which was soon replaced with the stronger and more reliable D-50. Within a couple of years, the 668 evolved into the BM Volvo 868, initially a forwarder designed for long hauls which was later equipped with a boom-mounted falling head and a clam bunk for the export market. It was by far the best-selling forest machine in the Nordic countries for many years, with close to eighteen hundred units built by the time manufacturing came to an end in the mid-1970s. A more significant role for the 868 was as a chassis for a feller-skidder. A felling head was mounted on a hydraulic boom. It severed trees and loaded them into large hydraulic skidding tongs without delimbing them. With this machine, Sweden had acquired its domestically built feller-skidder, the first of a long, illustrious line of mass-produced Nordic harvesters.

> “
>
> *In the late 1960s, I received a grant from the Finnish Natural Resources Research Institute. I used it to start developing a delimbing machine from a farm tractor. It cut the tree with a guillotine, and it had a radio-controlled winch. The Institute urged me to sell my invention to Valmet, but I didn't consent. But I did start to build it in partnership with Valmet. When the machine was ready, the guys from Valmet drove it out of the proto-shop into the yard to wait for the following day's test drives. During the night it started rolling down the hill by itself, and felled three trees along the way. It was a very eager machine.*
>
> Sakari Pinomäki

In 1969, at a small shop in Rottne, a country boy named Börje Karlsson had been manufacturing forest equipment for several years. He had listened to the concerns and ideas of loggers, and using a Ford tractor had developed a six-wheel forwarder. In spite of its country origins, it was a smart looking machine painted Ford blue. When the Rottne forwarder was introduced at the Elmia forest machine exhibition, one of the visitors came up with a name for it — “Blondi”. The idea for the name came from the fact that a new Brunett was being introduced at the exhibition. The machines were so similar that it was necessary to distinguish them with a blond joke. The Rottne Blondi became one of the most popular machines of its time, and was widely exported. Karlsson claimed it attracted just as much attention in the countries to which it was exported as did blond Swedish girls. “Do it in the forest with a Blondi,” became a company sales slogan.

By the end of the decade, the Swedish forest-machine business was the scene of frenzied development. New forest tractors were emerging regularly, such as the Drivax, Kockums 826, Robur Flexor and Hemek All-drive. Valmet also tried to squeeze its Finnish skidder into the Swedish market, with little success. Research had been conducted to determine the impact of thinning on forest growth, and the favorable results launched development of the first fellers for thinning. Ösa, MoDo Mekan and Storebo Bruk began work on tree-planting machines, producing prototypes and a few production models. Their design work consumed enormous amounts of money, but there was great faith in the future. Scarifying and harrowing of forests increased, leading to the development of stronger machines, which the Swedes sold to Finland as forest drainage machines called Godans (Frogs) because they worked in bogs. Meanwhile, debarking machines had moved back to the saw mills where debarking could be done more efficiently. They were replaced in the woods with chipping machines to utilize logging waste for energy production.

Competition increased and led to some restructuring. Lars Bruun's cooperative agreement with the VSA and Ösa was discontinued and he sold the production rights for his wheeled loader and processor to Kockums. Soon after, the Kockums Bruunett appeared on the market.

Kockums was now fully committed to the forest machine market, manufacturing rubber-tired skidders and feller bunchers for full-tree logging, and processors and forwarders for cut-to-length work.

By the mid-1960s, Finland also was recuperating from the recession. Imports, although still expensive, were increasing, and the Finns started buying forest equipment from Sweden — Hiabs, Ford Brunetts, half-track Boxers, Bears and Kalles. The most popular machines were Volvo's Little Bear and the Volvo 868, hundreds of which were imported. Some farm-tractor-based forest machines — Fords, Fiats and SMV Drivaxes — were also brought in, along with John Deere, Clark, Kockums and Volvo skidders. Even so, through the late 1960s, horses still skidded most of the logs to roadside. Finnish loggers did not like machines; they were awkward to use and often broke down. The short loaders caused extra work, and pulpwood had to be piled along the logging roads. But in spite of loggers' opposition, trainers began to appear at work sites to talk about work safety, data processing machines and other new concepts. Logging was changing to team work; now, the needs of the entire logging system had to be taken into consideration. At first, the horsemen laughed at these talks and at the machine men who were always covered with black grease. One veteran horseman taught his horse to snort from both ends whenever it passed a broken machine. But the machines had come to stay.

A program to drain swamps was started in Finland and excavators were brought in to dig ditches. Reforestation became more common, and the scarifying and harrowing it required created a need for even more machines. Techniques were devised for adapting the machines to the tasks they faced, and taught to the loggers. For example, the Central Forest Association Tapio provided instructions on how to free a machine that had sunk in a swamp by attaching a log to the wheels with a chain. As the wheels were slowly rotated, the machine climbed onto the log. Imported machines were tested on a variety of tasks, including a short-lived attempt to delimb trees with the dozer blade of a Timberjack skidder.

For the Finns, it was no longer enough that a forest machine was built in a Nordic country; now it had to be a Finnish machine. Consequently, the people at the Valmet tractor factory were disappointed when their first Finnish-made skidder did not immediately catch on. In 1964, Valmet, together with the National Board of Forestry and the Sisu truck factory, began designing a wheeled machine with frame steering and a trailer powered with Sisu hydraulic powered wheels. The prototype was introduced in 1965 as Model SMV-1, the letters referring to the three parties involved in its development. Although a Värtsilä loader was added, the machine did not progress beyond the experimental stage. In 1966, Valmet produced the Terra AK, equipped with a two-man cab because the engineers did not believe the operator could manage without an assistant. Tests later revealed one man could both run the machine and load it. The following year, the factory introduced the Terra 865 BK, a prototype skidder able to carry a short-log load. Soon after, a six-wheeled version of the 865 was built which incorporated an axle developed by Lauri Marttiini, a Rovaniemi knife maker who developed a business modifying forest machines for conditions in Lapland. By 1968, Valmet was being taken seriously as a forest-machine manufacturer. It had a variety of products to offer Finnish loggers, including two skidders, the 880S and the 1280S. They were followed a few years later by the 880K, a forwarder with a Cranab loader mounted on its roof. After several years of development work on forest machines, Valmet was now generating more than design costs and overtime at the prototype shop.

Valmet had begun as a state-owned company manufacturing ships, railway engines, airplanes and agricultural machines. It had a clear motive to enter the forest machine market: to provide the domestic forest industry with its own equipment. The next Finnish company to enter the business had different motives. Lokomo, in Tampere, was originally a steel foundry that went on to manufacture steam engines. It became a competent machine builder and during the war it made aircraft parts and barrels for rifles and cannons. By the 1960s it made diesel engines, excavators, graders, crushers and vibrating rollers. But the era of the diesel engine was coming to an end with the electrification of railways, the paving of highways and the growing obsolescence of other diesel-powered machines. Lokomo was searching for new products to build and forest machines looked promising.

In 1966, Lokomo signed a licensing agreement with Beloit in the US to manufacture logging and timber processing equipment. Lokomo began design work on a complete line of forest machinery, covering the harvesting and processing of

“

*We first logged with tractors in Ladysmith because we had a big blow down of timber. We had 12 tractors, D8 Caterpillars, hauling logs. We had a little A-frame deal to load with. Part of the idea of this operation was to develop pneumatic tires and log with trucks. It had been done a little bit where you only put two or three logs on your truck because they all had hard tires, hard rubber. For several years that was our biggest difficulty, pneumatic tires. They kept blowing out, in spite of everything we tried. The tire companies had men there, engineers, right along riding in the trucks watching them, taking temperatures. They'd get so hot they'd blow out. They increased the sidewalls, you know, and apparently that was worse, because they retained the heat. Then they went back to thinner walls. They tried every kind of fiber they could possibly find, even the steel belted that they rave so much about now. They tried that in 1937, built right into the tires. It looked beautiful. It wasn't a straight piece of steel. It was webbed, about six inches wide. It just looked marvelous, but in no time at all it was sticking out through the rubber in chunks, and all broke up. The Michelin tire people did that. Goodyear and Firestone were also there, but Goodyear turned out the best tire and spent the most money. Marvelous. I don't think Goodyear made a nickel there for two or three years, but they finally got the answer.*

*You can't complain now. They've got the right kind of air pressure, and right kind of width of a tire, all that sort of thing. If you blow out tires today, its pretty well your own fault — overloading or letting the air down too low and you're not looking after them. Of course they never could log these mountains the way they are now without the pneumatic tire. It would cost too much money to put a railroad up there.*

Joe Cliffe

timber from stump to mill. It also signed an agreement with Hankkija, a cooperative which specialized in importing agricultural and forestry equipment, to provide a sales and service network. The first machine built was a skidder based on a Beloit model, introduced in 1968. But the market was full of Timberjacks and Kockums, and Finland was also moving from full-tree skidding to forwarding of processed logs. In response, Lokomo brought out the six-wheeled Teli-Lokkeri forwarder the following year. In appearance it was radically different from its competitors. The cab was mounted in front of the engine, providing excellent visibility but a rough ride for the operator. It had a 135-horsepower Perkins engine, an Allison two-speed power-shift transmission, a Cranab loader and a capacity of fifteen tons.

A third pioneer in the new wave of Finnish forest machines was Sakari Pinomäki, who had entered the business a few years earlier with the Pino-Teho debarking machine. When debarking moved to the mills, he started developing a multi-purpose machine and two years later, in 1968, a prototype of the Pika 50 was ready. It consisted of a modified Valmet Terra 865BK tractor which felled, delimbed, bucked and bunched trees. Working in cooperation with the Natural Resources Research Institute, the Work Efficiency Association and Valmet, Pinomäki built 60 machines — a few of which were exported to China — before production was terminated.

The ability of the Finns to build the required machines had now been demonstrated, at least to their own satisfaction, and the need for mechanization was steadily becoming more obvious. In 1966, Jaakko Piironen, managing director of Metsäpalvelu Oy, published an article noting that, on average, a mechanized logger in Sweden produced five cubic meters of wood per day, and that North American loggers achieved similar results. But in Finland, loggers averaged less than two cubic meters of wood a day. Piironen warned that if Finns continued to rely on imported solutions to their logging problems, the gap in productivity would continue to grow. His message did not fall on deaf ears.

When Kockums introduced the Logma delimbing machine in Sweden in 1970, Domän placed an order for 125 of them.

The Pika 75, was introduced in 1972. Equipped with a chain saw for felling and bucking, and delimbing knives, it was one of the first Finnish harvesters.

The Ösa 670 feller buncher had a hydrostatic power transmission and a unique levelling cab.

Ösa's prototype processor, the S-3 had two cabs, one for driving the base machine and the other for operating the processor.

The Ösa 710 processor could be converted to a harvester by outfitting it with a 750 felling head. The base machine was the Ösa 260. Now only two machines were needed in a mechanized timber harvesting system. One machine felled, delimbed and bucked, the other transported logs to the roadside. Ösa was flooded with orders, and production lines were overwhelmed.

Lokomo's first processor, the 961, had pockets for different log lengths. When the pockets were full, the 961 dropped them into a neat pile.

Lokomo's first harvester, the 960, felled, delimbed and bucked trees up to 60 cm in diameter. A version of the machine was later made for the export market, but sales never took off.

Felling heads in the 1970s were huge. The Ösa 640 felling head utilized a chain saw, replacing shears which shattered the wood, especially when it was frozen.

The delimbing chains and feed rollers of the Ösa 710 processor. The spikes on the rollers made deep marks into the logs, sparking a major research project in Sweden. The solution was found in the form of rounded, chisel-shaped roller teeth.

The most successful forest tractor of its time, the Volvo 868, with an Ösa's 710 processor installed on it.

Kockums' three-machine timber harvesting system from the 1970s: feller-buncher, processor and forwarder.

The Franklin Prehauler, built in Virginia, was a prominent example of a new generation of North American forwarders.

The Ponsse PAZ, which first appeared in 1971, was built at a rate of five machines per year for contractors in the vicinity of its Finnish plant. “Whenever we had a machine ready, a buyer lined up,” recalled Ponsse’s Einari Vidgrén 25 years later.

Ösa manufactured nearly two thousand Volvo 868 forwarders at its Alfta factory from components supplied by Volvo. Ösa had the overall responsibility of machine design.

The basic concept of a Scandinavian forwarder emerged in the 1970s: Six or eight wheels, frame steering, a hydraulic loader and an engine with good torque. Since then, improvements have been made, but the basic concept has remained the same. Pictured is a mid-sized Valmet 886.

The Gafner Iron Mule, first built in 1957, went through several improved models and was one of the most successful pulpwood forwarders ever made.

The Log All feller skidder, developed in 1968 by Logging Research Associates in Quebec, had a felling-loading head mounted on a knuckleboom. In combination with the LRA's roadside processor, the Arbomatik, it provided the world's first fully mechanized logging system.

Timberjack's pulpwood forwarder, the Pulpjack, followed on the success of its skidder, but was a short-lived machine.

The Serlachius company experimented with a container system called Serla at the beginning of the 1970s. The intention was to eliminate the re-loading phase at roadside. Forest tractors loaded up trailers in the forests, and pulled them to the road-side where they were hitched to highway trucks.

A Timberjack skidder working at the Sinarajah Mechanized Logging Project in Sri Lanka in 1974.

A US-built FMC tracked skidder. This high-speed, low-ground-pressure machine first appeared in 1974, and today is manufactured by Kootenay Equipment in British Columbia.

Timberjack brought out this model 480 tracked grapple skidder in the early 1990s in response to a need for a ground skidding machine which could operate on steep slopes with minimal site disturbance. It could work on 60 percent slopes.

The Reforester tree-planting machine was developed at the University of British Columbia in the late 1970s. Seedlings grown in special containers were planted with the manoeuvrable hydraulic booms.

While the 1970s was the decade of giants, a counter culture emerged and machine builders began developing smaller machines. The operator of the Swedish Forest Ant walked alongside the machine and steered it like a dog on a leash.

The Finnish-built Makeri harvester, developed by Muisto Laine for thinning and harvesting pulpwood, was mounted on a Bobcat. The felling head had guillotine blades and turned the tree to a horizontal position for cutting off the top. The Makeri was moderately successful in Finland and some are still operating on plantations in Spain.

# The Return of the Processors

The use of forest machinery in Sweden escalated dramatically at the turn of the decade. In 1969 there were 1,200 horses working in Swedish forests, but by 1971 there were only four hundred. During the same period the number of farm tractors used in the woods dropped by more than fifty percent. The number of feller-skidders increased from four to more than sixty, and the use of other purpose-built forest machines climbed just as steeply. In 1972, the Skogsarbeten (Forest Operations Institute of Sweden, established in 1964) predicted that by the end of the decade mechanized felling and delimbing would increase from five percent to seventy-five percent of the timber harvested, and that the cut-to-length method would become the most widely used system. In fact, it reached a usage of more than eighty percent, and full-tree logging methods were used only on rare occasions.

As it turned out, Skogsarbeten's predictions were not unrealistic. At the time they were made, a few processors and felling machines were already at work and their commercial prospects looked good. Although product development costs at the factories were still in the red, markets were growing explosively. There was a great deal of talk around board-room tables of export opportunities, along with the expectation that big machines would yield higher profits than small ones. Equipment manufacturers were frantically hiring young talent for their engineering departments.

In the early 1970s, equipment sales began to pick up as new delimbing machines, feller-skidders, feller bunchers and processors appeared on the market. Kockums introduced a new processor in early 1970 — an improved model of the Kockums Bruunett — and a delimbing machine which sold very well. A single forest company, Domän, placed an order for 125 Logma delimbers. Brånemo introduced an extremely efficient delimbing machine, while Morenius produced a new falling machine.

After machine builder Lars Bruun and Kockums terminated their working agreement because of differing views about the Kockums Bruunett processor, Bruun established his own company in Filipstad, Bruun Systems. He soon shocked the industry with the announcement that he was developing machines for Kockums' biggest competitor, Volvo. In 1971, the new partnership's first product was unveiled, the Volvo BM 880 processor. A couple of years later a bigger version, the Volvo BM 980, appeared, followed in 1975 by the even bigger and more efficient BM 995. Similar in function to some of the earlier North American harvesters, these machines were equipped with a felling head that cut the tree and moved it to a processing unit which delimbed, bucked it into logs and, in the pulpwood processor, sorted the logs by species and piled them in one of several bins. When a bin was full, the logs were dropped for retrieval by a forwarder.

With thinning harvests on the increase, Skogsarbeten conducted a series of tests which demonstrated that costs decreased by about one-third when the distance between logging trails was increased from twenty meters to thirty meters. In response, Jonsereds, a small machine shop which expanded to build knucklebooms in the wake of Hiab's success, developed ten- and fifteen-meter Cranab booms to carry felling heads on light falling machines and loading grapples on forwarders.

Ösa was also running at full speed. In 1971 it expanded its factories and introduced its first processor, the Ösa 710. The prototype was a wild-looking pile of iron, but the production model looked more convincing. It had two cabs, one for driving the base machine, the other for operating the grapple. The processor head incorporated a relay-controlled measuring system designed to calculate the optimum use for each tree. A refined version of the measuring system developed a decade and a half earlier by Rudy Vit, and then ignored, the Ösa system was the first European development of this technology. It was devised by a woman, Stina Mickelsson, who was hailed in Sweden's media as "the world's first processoress."

Ösa's processor gained added fame when Domän installed an air-conditioned cab on one of its machines, the first such improvement in the Nordic countries. The air-conditioning equipment and its installer came from Germany, at a reported cost of 10,000 krona. Because so few drivers were qualified to operate these new machines, they were becoming respected professionals and treated accordingly; air-conditioned cabs were considered a measure of this respect. At the same time, Ösa also introduced a four-wheel forwarder, the SM-462, designed for maneuverability. It could be equipped with half-tracks to redistribute its weight and reduce damage to the terrain. The new machines, particularly the processors, required more from the operator than just a tractor operator's license and a heavy foot on the gas pedal, so Ösa established its own training center at Alfta, Sweden, with Lasse Larsson as its head. In 1973, the company brought out its new Ösa 670 feller-buncher with vastly improved hydraulic systems, a hydrostatic transmission and the industry's first self-levelling cab to improve the operator's performance. "Even in hilly terrain, the operator sits like he would at home in his armchair, and now he can really concentrate on his work," the company boasted. The same year, 1973, it produced Sweden's first harvester, the Ösa 710, that felled, delimbed, bucked and carried cut-to-length logs to a point where they could be taken by a forwarder. Even with two cabs and a weight of more than twenty tons, it was a significant evolutionary step. The modern mechanized timber-harvesting system had come together in only two machines: the harvester and the forwarder.

Perhaps Ösa's success made it believe it had the Midas touch, able to turn everything it touched into gold. Or maybe it was in a hurry to respond to the competitive challenge posed by Volvo, as well as Kockums, which bought Stensele Mekaniska in 1973. Both companies were well ahead of Ösa in sales. Whatever the reason, Ösa set out to develop a stump extractor. It built an enormous stump-lifting grapple, which pulled stumps from the ground like they were carrots. The company also experimented with a root-cutting machine with large shovel-shaped blades that

"

*We introduced the first slasher Omark built in 1966, and my first job with the company was to ramrod the promotion of the sale of that little slasher which, we called the Sawbuck. It was a little machine, it was very primitive compared to today's slashers. It was operated from electrical controls out of the cab of the loader, and evolved into the very sophisticated hydraulic machines that we have today.*

*The slasher was controlled from the hydraulic loader. We were actually ahead of our time. We had trouble selling the slashers because people just didn't understand what a cost savings they would be.*

*Slashers were not as popular then because there was lots of manpower. We had plenty of men with chain saws. Slashers are used to reduce big, long trees into short pieces, long logs into saw logs and long stems into pulpwood lengths. At that time, we had plenty of manpower and there were not as strict workmans' compensation laws. People weren't concerned about their costs as much, it didn't cost as much then to maintain a crew.*

*But as the forestry market became more sophisticated, the compensation costs escalated, and the need for safety was becoming more obvious, loggers became more concerned about reducing the costs of producing pulpwood and timber. A slasher eliminated all the men with chain saws on the deck, or the landing. It eliminated that safety hazard. It eliminated all the extra men who were usually not well trained, for whom the accident rates were pretty high.*

*Then a knowledgeable loader operator controlled everything on his landing. He didn't rely on people on the deck to make those decisions. The people who did the bucking by chain saws were very careless at times. If a sawmill wanted a 16-foot 3-inch log, sometimes those logs would come in at 15-foot 9-inches. So the mills weren't getting what they wanted. Instead of a 16-foot length they had to cut back to a 14-foot log and it was a waste.*

Mal Crawford

wrapped around the tree and pushed down into the ground, cutting the roots. A grapple lifted the whole tree out of the ground. Both machines produced additional wood, but it was contaminated with rocks and dirt that dulled chipper blades. The ideas were theoretically sound, but mills were unable to overcome the problem of foreign material and the method proved too expensive. The venture was a major setback for Ösa.

Until the late 1960s the stream of forest machines flowed from North America to Europe. In the early 1970s, the Nordic manufacturers came to the conclusion it was time to reverse the current. They were sure their machines and logging methods were superior to the North Americans' and were convinced they would sweep the market. Because the major Nordic manufacturers kept a close eye on each other, they all arrived on the American and Canadian markets at about the same time. At first, the Nordic machines aroused a great deal of interest. All the companies made a few sales, which encouraged additional investments in the export venture. Lokomo delivered dozens of machines to Canada, which inspired it to acquire a half-interest in a Quebec machine shop where its machines would be assembled to avoid import duties. But conquering the North American market proved to be more difficult than imagined. Encountering stiff competition to their invasion, the Scandinavian contingent was soon in disarray. Poor choices in partners and distributors were made, and they neglected to establish service and spare parts networks at the outset. The disastrous venture was summed up by Lokomo's chief engineer, Matti Kaarnametsä: "It was a practical lesson in how not to internationalize. The hollowness of the journey started to make itself clear when I was sitting in a logging camp a hundred kilometers north of Quebec, and the machines standing around me were broken. After the mid-1970s, we all went home badly beaten."

The rapid mechanization that went on in North America during the 1960s and 1970s was accompanied by significant structural changes in the forest industry. Which one caused the other is an open question, but the result was a shift in ownership of logging operations in much of North America. At the beginning of the mechanization period, the prevalent organizational mode in eastern Canada and the American South included a jobber or dealer who served as a middleman. This person usually contracted with the timber-using companies for the delivery of a certain volume of pulp wood. He then hired or assembled one or more logging crews to obtain and deliver the wood, usually by river drive in the northern forests or to a railway siding in the South. A key role of the jobber was to provide interim financing to the logging contractor. Since most of the work entailed hand labor, these up-front costs were not very high.

With the onset of mechanization, this organizational structure experienced considerable stress. Initially, much of the new equipment was financed by dealers, who could handle the cost of a three-year-old Coca Cola truck, a used Cletrac crawler and a new wheel saw. When the equipment began to include items such as a Logger's Dream, the situation became more problematic. And when it got around to a new Timberjack or a Franklin skidder, the system fell apart. It might take a logger two or three years to pay off a skidder, and dealers were not inclined to carry for that long someone as marginally solvent as most loggers were in those days.

“

*John Hingman was the President of International Paper Company in Canada and was active in the Canadian Pulp and Paper Association. He was impressed with the work done by Logging Research Associates (a five-year joint venture begun by the Canadian International Paper and Quebec North Shore Paper companies in 1962) and started talking about doing something like this in the U.S. Then Ken Ralston, who was with the American Pulpwood Association, talked with Hingman about funding research in the U.S., primarily in the South. When they got that together, they asked me if I would head it up. I was with Hingman in a hotel one night and he asked me, "How would you like to leave Bowater for five years and head up this organization?"*

*I thought I'd died and gone to heaven. I had a logging engineer, a mechanical engineer, an industrial engineer, and a forester. Well, they were all foresters, but these were their primary jobs. So we set up the Harvesting Research Project and were funded by six companies originally. Our budget was about $50,000 a year. We had an office in Atlanta and flew all over — sent people to Sweden and Finland and Canada to look at all the equipment. When we came back we started trying to figure out what we might be able to do and what the problems were.*

Tom Walbridge

Nor did the system work for the dealers themselves, or for the timber companies, if they bought the equipment and charged loggers to use it. If the owner of the machine was not the operator, or directly in control of the operator, the machine invariably was mistreated and broke down frequently.

The dealer/jobber system broke up in two directions. Either timber companies set up their own logging operations, which was an option extensively pursued in eastern Canada, or the mills bypassed the dealers and financed logging contractors directly, which occurred more frequently in the American South.

One of the requirements of the mechanization process, as the Swedes were learning, was the development of skilled machine operators. It is unlikely this would have come about under the dealer/jobber system. With the emergence of owner-operators who had a direct interest in the upkeep of their machines, the professional logger was born. To get a job running one of the new skidders, loaders or feller bunchers a man had to develop skills and take pride in his work. The emergence of a more skilled workforce, in turn, speeded up development of more highly mechanized equipment.

Another significant contributor to the mechanization process appeared on the scene in 1968 when six US pulp and paper companies, in cooperation with the American Pulpwood Association, set up the Harvesting Research Project in Atlanta, Georgia. The companies loaned engineers to the project on a full-time basis for five years under the leadership of Tom Walbridge, a Bowater Timber forest engineer from Tennessee. The group looked at what equipment was available in North America and Europe, talked with loggers and focused on developing ideas about the next stage in the mechanization process. One of the first things to evolve out of this group was a small feller-buncher head to mount on a Bobcat, designed by Joe Blonski, the group's mechanical engineer. When the project's funding ran out, it was relocated as a research facility at Virginia Polytechnic Institute in Blacksburg.

Running parallel to the rapid evolution of rubber-tired equipment was the short-term use of cable yarding systems in eastern Canada and, to a lesser extent, in parts of the northeastern US. Prior to World War Two there was scant use of cable yarders outside the Pacific Northwest and the southern US. In the immediate post-war period, faced with an urgent need to mechanize, eastern Canadian pulp-wood loggers turned to cable yarders as a readily available form of mechanized equipment.

One of the first cable yarders put into the eastern woods at this time was a Lawrence swing yarder purchased by Canadian International Paper (CIP) for its Clova, Quebec, operations. It was this machine that brought Bob Symons to eastern Canada, where he stayed to help set up the company that would become Timberjack. Several other logging companies began cable yarding through the late 1940s and early 1950s, developing techniques of yarding bundled pulp wood.

In 1950, CIP bought another Lawrence yarder to work in the mountainous area in the Gaspe region of Quebec. It was set up and operated by Bill Mosher. He was a member of an extended family of innovative West Coast loggers whose ancestors had begun logging in Rhode Island in the early seventeenth century. They moved to Nova Scotia in 1676, where they logged and milled lumber. In 1900, most of the loggers in the family rented a boxcar, loaded up their possessions, wives, children and logging equipment, and moved to British Columbia, where they started over again, handlogging. Ever since, there have been Moshers working in logging operations throughout North America, often where a new type of equipment is under development.

Many of the eastern yarders were custom built by Timberland Machines, which by 1961 had sold twenty-two yarders to just one company, Abitibi Pulp and Power. The most widely used model consisted of a sled-mounted timber A-frame about ten meters high, with a four-drum winch. Using a high-lead system, this rig could yard up to 200 meters. One company, Kimberly-Clark, used a rented Logger's Dream to move bundled pulp wood at Longlac, Ontario, and was so pleased with the results it built its own version of this southern favorite. Marathon Paper Mills, one of the most highly mechanized companies in this era, advanced to night yarding, with a battery of floodlights mounted on the top of the A-frame and chokermen wearing miners' headlamps.

> “
>
> *In the mid-1970s we took our pride of the day, the big Lokomo 961 harvester, to the Elmia Exhibition in Sweden. It had length-measuring automation. We had two models, one that cut logs short and one that cut them long. At Elmia we had the model that cut them short.*
>
> *The King of Sweden, Karl Gustaf, opened the exhibition and then toured the displays with the local forestry elite. The procedure was that when the King's delegation stopped at the display, we would demonstrate our machine. We had only three logs for the demonstration. The first log was chopped into short pieces like it was supposed to, with the pieces going nicely into the machine's pockets. But then the early automation went crazy. The next tree was a real nice piece of fir; the whole log went through without being cut. We started sweating, and the operator grabbed the third tree. It, too, went through. A royal mistake, and we were flustered. But then we started hearing what the delegation was saying. They didn't realize what had happened. They thought we had developed a machine that can cut logs both long and short.*
>
> JUKKA MOISANDER

With the appearance of skidders, cable yarders disappeared from the eastern Canadian pulp stands. Although the cable systems could move bundled wood to roadside at a low price, the cost of cutting and piling the shortwood pulp logs increased to the point where the whole operation became uneconomical. The only locations in eastern Canada where cable yarding survived through the 1970s were Newfoundland and New Brunswick, where a new generation of small cable machines was introduced to yard the small trees found there. One of the most popular was the Ecologger, a lightweight, portable cable machine built by Rosedale Equipment in British Columbia.

In the late 1970s, Bowaters began using a cable system developed in Scotland, consisting of a short mobile tower with a four-drum winch that included a skyline setup. Using standing trees for intermediate support, it yarded at high speeds over a range up to 450 meters. These machines — the Smith Timbermaster and the Iglund-Jones Mini-Alp were two examples — used external power sources, such as a farm tractor's power take-off, and were designed for thinning. In 1980, Norman Williams, a Timberjack dealer in Courtenay, British Columbia, and local logger Bob Woods mounted a Mini-Alp in the rear of a Timberjack skidder to thin second-growth Douglas fir. Skylead Logging Equipment in Enderby, British Columbia, still manufactures a modified version of this configuration for mounting on a wide range of carriers. They are used for logging steep, wet or environmentally sensitive sites around the world.

Several other manufacturers of small cable yarders are scattered around the world, all producing machines designed primarily for thinning, but also used on small clearcuts. Increasingly, they are used on steep or wet ground, where rubber-tired and tracked machines are unable to operate. Koller, in Oregon, is the oldest North American builder and markets several models. Urus yarders are built by Hinteregger in Austria and South Africa. Owren Yarding Systems, of Norway, has been in the business for 25 years and produces a hydrostatically driven yarder mounted on a forwarder chassis. Several Japanese manufacturers, including Nichimen, Iwafuji and Ryoshin, build small, mobile cable yarding systems. Being relatively low-cost machines, the small cable yarders are increasingly used around the world, wherever conditions warrant.

Typically, these cable yarders are capable of utilizing standard high-lead and skyline systems, as well as a range of new self-propelled carriages. Compact units containing small gas or diesel engines and winches, these remote-controlled carriages travel along a single, fixed skyline and lower a skidding line to haul in logs. Manufacturers include Mäki, from Finland, and TLD Gauthier from Quebec. The latter's Telecarrier can operate over 100 percent slopes at distances of 500 meters, carrying a 3,000 kilogram payload.

“

*We set out to conquer Canada in the winter of 1974. We loaded a processor, the BM Volvo 880, onto a ship in Sweden, and unloaded it at Montreal. The machine was driven from the port to a local machine shop where it was loaded onto a trailer.*

*Then we started pulling the trailer hundreds of kilometers through the night. The temperature was close to –30 degrees Celsius. Encountering the first telephone line, we discovered that the machine was too high, it wouldn't pass under the lines. We fetched a cluster of young spruce saplings from the forest and built flexible antennas on top of the machine; the lines would slide along the antennas and over the machine. The roads were deserted, and we drove at a good speed through the freezing cold winter night. Arriving at our destination at dawn, we saw what destruction we had caused. Tangled around the antennas were a good-sized bunch of Canada's telephone lines.*

*The cold ride had been too much for the machine. It seemed like it was in a deep-freeze, and it just wouldn't start. We borrowed an industrial heater from a construction site in a neighboring village, covered the machine with a tarp and blew hot air under it the whole day. By that evening the machine had defrosted and we were able to start it. Then we drove it to the test area, where we showed the Canadians just what kind of machines the Swedes could build.*

GÖSTA GUSTAFSSON

By 1970, logging in the Pacific Northwest had been completely transformed from its pre-war condition. There was no steam machinery left, wooden spars were gone, and railways were reduced almost to the point of extinction. As logging of smaller second-growth timber commenced in Washington and Oregon, skidders were heavily used. In old growth stands, they were occasionally used as intermediate haulers between a mobile steel spar and a landing or dump.

Perhaps the major innovation of the era, especially in coastal British Columbia, was the widespread adoption of grapple yarders. Their appeal was direct and simple; with the addition of an inexpensive grapple to an existing cable yarder system, two to four workers could be eliminated. In its basic form, the grapple yarder required interlocking mainline and haulback drums, plus another drum to operate the grapple. With the appearance of excavator chassis equipped with turntables, the machine evolved into a swing yarder which could pull in a turn of logs, maneuvering around obstacles, and pile them at roadside for easier loading. The addition of another machine, such as a crawler tractor, as a mobile tail hold speeded up operations even more.

The first swing yarders were developed by Bill Wainwright in 1963 at Crown Zellerbach's Nitnat operations on Vancouver Island, in cooperation with Skagit. An SJ-7 loader was modified with the addition of a 12-meter boom, and could yard up to 150 meters. These machines were not widely used in the US, probably because of the extensive network of roads they required, taking a lot of forest land out of production. In British Columbia, where almost all the forest land is publicly owned, the cost of building these roads was deducted from stumpage fees, which was not possible in the US. As a consequence, during the 1970s and early 1980s, when most British Columbia coastal loggers switched to grapple yarders, their skyline equipment ended up in the US, only to be repurchased a decade or so later, when grapple yarders began to lose favor.

The mid-1960s also saw the beginning of aerial logging on the West Coast. MacMillan Bloedel used two surplus military balloons at Sproat Lake in 1965 to provide lift on a high-lead cable system. The following year forest engineer Chester Matheson used a 3,000-cubic-meter balloon, controlled by a custom-built Skylift winch system built by Frank Lawrence, to log in the Vancouver city watershed. Both these systems failed, primarily because of inadequate balloons.

In the 1970s, a consortium of Canadian and US interests set up a company called Aero Lift to develop a free-flying heavy-lift logging balloon called the Cyclo-Crane. It was a powered aircraft which achieved additional buoyancy by rotating. A prototype model was destroyed in a 1982 windstorm, stalling further development. In the late 1980s, Faye Stewart of Bohemia Lumber in Oregon, along with Frank Mosher of British Columbia, developed a new system using a custom yarder built by Washington Iron Works. It used a 525-horsepower Detroit diesel on a two-drum winch to hold down a balloon with 10,000 kilograms of lift that could carry a turn of logs completely off the ground. Two of these machines were purchased by Skyhook Enterprises of Campbell River, British Columbia, and five years later were still operating successfully.

The use of a more conventional aircraft to yard logs, the helicopter, has long been the dream of woodland managers with too much tough ground and too small budgets for road building. Consolidated Paper of Quebec was the first North American company to try a helicopter for logging when it used a Sikorsky to fly bundles of pulpwood in 1957. Two years later, Anglo-Canadian Pulp and Paper undertook some trials at Forestville, Quebec, using a Sikorsky S-55, and confirmed what was already suspected — that it was too expensive for moving high-weight, low-value materials such as logs. In 1975, British Columbia's first heli-logging operation was launched on Redonda Island by logging contractor Ian Mahood, forest engineer T.M. Thomson and Okanagan Helicopters, using a Bell 214, which proved to be too light for the job.

The first really successful helicopter logger on the continent was probably Jack Erickson, a California logger. After developing his system in the US, Erickson set up Silver Grizzly Logging on the northern British Columbia coast in 1977, opening the way for a generation of heli-logging companies. Within a decade there were a dozen helicopter logging contractors in business, and one major forest company, International Forest Products, with its own heli-logging division. The most popular aircraft are Sikorskys and Boeing Vertols. In 1991, Vancouver Island Helicopters began using a Russian-made Kamov, which performed exceptionally well as a logging machine. By 1996 almost 20 percent of British Columbia's coastal timber harvest was logged with helicopters.

On the rest of the continent, at about the time Tom Walbridge's research group was established in 1968, the next stage of the mechanization process was underway. By 1970, the concept of moving short lengths of pulp wood from the stump to roadside had been more or less abandoned. The most prevalent system involved the use of a rubber-tired skidder to skid tree-length logs or whole trees to roadside where they were processed into logs. This development called for a range of machines to process the logs into a form most easily transported to the mill.

A basic principle of long-distance hauling is that the more uniform the piece size, the lower the cost of handling and transport. Theoretically, if the end product is feed stock for a pulp mill, the least expensive form in which timber can be moved is as a wood chip. At this time, because of a lack of chipping machines capable of producing clean, high quality pulp chips, there was a demand for equipment to convert the trees — which, increasingly, included limbs, tops and needles or leaves — into intermediate products. Much of this equipment evolved during the 1970s. There were essentially four steps in this process. Trees needed to be delimbed, debarked, cut to the desired log length, and reduced to chips. Since virtually all of this work was done at roadside or a landing, the machines only had to be mobile enough to get over a logging road into the forest.

> *In 1976, Queen Elizabeth visited Tehdaspuu's work site in Toivakka. Taking part in the Queen's forest tour was a varied group of Finnish and foreign journalists of both sexes. Quite a few of the Finns already had a "buzz" on. At the forest sites arranged by the Central Finland zone of supply, there was a new forest machine called the Pika-75 (Fast-75). The machine grabbed the base of a big, thick-branched spruce, delimbed the branches up the tree to a height of about 7 or 8 meters while the tree was still standing, and then felled the tree and continued to process it into various timber products. A female journalist watching the machine work remarked with astonishment: "It's a terrible machine. First it rolls up the hem, and then after that it fells it." A voice from behind said: "Yeah, it works like a Finnish lumberjack."*
>
> From the book "Tehdaspuun aikaan ja paljon ennenkin" (The Tehdaspuu era and before)

The removal of limbs and tops, until well after the development of the chain saw, was done with axes. The first power saws, including chain saws, were not well suited to delimbing trees, until the introduction of the lightweight, one-man saw. Even then it was a slow, dangerous procedure. As other phases of the logging cycle became mechanized, delimbing became a significant problem, particularly with the move to whole-tree skidding. The volume of timber a grapple skidder, working behind a feller buncher, could deposit at the landing presented an enormous problem for workers delimbing with power saws. Not only did it require a large workforce, but the men wielding the saws had to work on the ground among fast-moving machines. The normal hazards of the task were increased substantially.

There were some attempts — in Ontario during the mid-1950s, for instance — to utilize chain flails for delimbing. Originally developed as a debarking device in Sweden, the chain flail was employed in an experimental delimber built by Clark Horncastle for the Ontario Paper Company, but it did not progress much beyond that. It was a roller-fed machine, with two flails on the inside, through which the trees were fed.

In the South, with the adoption of grapple skidders, a simple device effective on pine limbs came into wide use. The delimbing gate was a vertical, welded pipe grid, firmly anchored. As a turn of logs was backed through the grid, tops first, most of the limbs were removed. In 1972, a different approach to flail delimbing was devised by Gordon Franklin, equipment manager for Manitoba Forestry Resources Ltd. at The Pas. His flail was mounted on the front of a skidder, which drove over several trees at once, removing the limbs. Several commercial versions of this delimber were produced over the next five years, but it was never a satisfactory machine as it usually left a few limbs on the trees.

In 1977, the first stick or stroke delimbers appeared from Sweden, utilizing the same principle as some of the harvesting machines developed twenty years earlier. The imported machines were very expensive, consisting of a complete unit mounted on a rubber-tired chassis. The tree was picked up by its top and delimbing knives, which folded around it, were propelled down the stem along a long stick or boom. Within a year or two, several eastern Canadian companies — Denis, Koehring, Harricana, Bourbeau, Roger — came out with less expensive versions of the stroke delimber, so that by the end of the decade it had almost replaced the flail delimber. Still, according to a 1993 *Timber Harvesting* magazine survey, 74 percent of US logging contractors used chain saws as their primary delimbing tool.

There was a parallel development of portable debarkers. Of course, debarkers of various types — most of them drum or ring machines — had been used for several decades in mills, but they were too large to consider taking into the woods. A lot of hardwood logs were hand-peeled in the woods, and some softwood pulp logs. The Swedes and Finns had been working on portable debarkers since World War Two. In Finland, the focus was on small, hand-held power tools to replace peeling spuds, none of which were very successful. In Sweden, research proceeded in another direction. A 1950 radio broadcast dealing with the debarking problem prompted Erland Andersson, a foreman from Vargön, to inform the engineers at VSA of a prototype debarker sitting in his wife's wash house. It consisted of a rotating drum with a set of adjustable debarking blades inside it. The VSA and the Billerud forest company built a working model which could debark a thousand logs a day. It attracted

“

*I started shortwooding in 1971, and we thinned the shortwood. Started out with little old bobtail trucks and a bow saw. Then later on, in about 1973 or 1974, the little old pre-haulers come out. Little Massey Ferguson Treevers they called them. We hauled a lot of shortwood out on those. Hauled with them until 1978 or 1979 when we went to treelength in the woods and using feller-buncher shears to lay it down, and grapple skidders and limbing gates.*

*First limbing gate I ever built or ever seen was when an old boy in Florida one night drawed out a diagram of how to build it on a beer tab. We was setting in a bar down there and he was a logger, but I didn't know it until we got to talking and he drawed out how to build this limbing gate. For years that was the way to knock the limbs off the trees here. Back in through it, and just knock the limbs off, then load them. 'Course now there's so many different kinds of delimbers you can't count them all.*

Tom Burch

attention from European and North American engineers, the latter flying a truck load of logs in from the US for a test. This machine eventually evolved into the Cambio drum debarker which was taken over and developed by Kockums. It was the father of an entire new generation of debarking machines.

By the late 1950s, several portable debarkers were available from manufacturers in Europe, the US and Canada. Some worked on the principle of passing the log through a ring of scrapers or knives which took off the bark; some used chain flails, or wire brushes, or a hand-held device of rotating knives. Apart from the Cambio, none of them was particularly successful, working only on some species of tree, or at certain times of the year but not others.

In 1959, Norval Morey, a Winn, Michigan, logger and sawmill operator, ran across an out-of-luck mechanic named Bob Baker who had a new concept for a debarker but no wherewithal to develop it. Morey put up the money for a patent, to build a prototype and to register a manufacturing company. One night in the local tavern they decided to incorporate their two names and call the firm Morbark. The Morbark debarker utilized a cylinder faced with a series of raised scrapers. Two powered roller wheels spiralled the log over the cylinder, scraping off all the bark — softwood or hardwood, frozen or not. They sold their first debarker to a local sawmill for $4,000, and immediately got an order for four more. During the remainder of that year they had $300,000 in sales, the next year over $1 million. A man debarking by hand could do a cord of wood a day in winter, maybe two cords in the summer. With the Morbark machine, two men could run through sixteen to eighteen cords a day.

Within a few years, Morbark customers started asking for a chipper to use along with their debarker and, after purchasing one from another supplier for a couple of years, Morey decided to build his own. One thing led to another, and within a few years Morbark was building a wide range of debarkers, chippers, grinders, peelers, sawmill equipment, tractors, feller bunchers and other wood-related equipment — all of it out of a plant at Winn, where Morey had grown up and begun logging with horses in the 1930s. By this time a number of other portable chippers were available in North America, the most popular being the ones built by Peterson Pacific in Oregon, owned by Nels Peterson, who had produced perhaps the first machine that could delimb and debark a log, and then convert it to chips.

Mechanized bucking began at a much earlier date with the skidding of tree-length logs by crawler tractors. This encouraged the introduction of a variety of slashers. The inclusion of other processing machines in the logging cycle increased the need for a single-function bucking machine. Delimbing with a chain flail or a stroke delimber could be a relatively inexpensive process with machines capable of moving along the stem of the tree at a rapid rate. Having the same machine make a cut across the tree slowed down production and raised costs considerably. In the early 1960s, a system known as bundle bucking was widely used at mill wood yards in the South. Gigantic chain saws mounted on a guillotine frame cut entire truck loads of tree-length logs into pulpwood lengths in a few passes. Perhaps the most popular of these bucking systems was the Currie Cost Cutter. Although the system was remarkably efficient, it was expensive and required special log trailers.

In the Great Lake States and eastern Canada a variety of slashers were produced through the 1960s and 1970s and gradually made their way south. Typical of these units, and perhaps the most popular, was one produced by Tidewater Manufacturing of Brunswick, Georgia. It consisted of a large circular saw which was fed tree-length logs by a knuckleboom loader, which also stacked the shortwood logs onto a truck or railcar at the rate of ten cords an hour. A similar machine, the Prentice Saw Buck, came on the market in 1968.

The next step was to develop machines which combined two or more functions. One of the first of these roadside processors was built by a Minnesota logging contractor, Ray Hahn. He began his working life as a teenager in the northwest corner of the state, logging pulp wood with horses. After the war he went back into the logging business, using Caterpillars to skid hardwood logs. About 1960, he moved to the north shore of Lake Superior, where a lot of pulp wood was mixed in with the hardwood. Finding it difficult to compete with other industries for workers to log pulp wood at wages he could afford to pay, Hahn began mechanizing as rapidly as possible, acquiring the first Drott feller buncher in the state, as well as the first grapple skidder, a Cancar Tree Farmer, to skid whole trees to the road. That's where the bottleneck occurred — using men with chain saws to limb and buck the trees into logs.

“

*I came to work for Volvo in 1969. At the time, the company had a rather aggressive regional manager who thought the South African market had enormous potential for forest machines. We decided to send one SM 868 forwarder there by ship, and fly me to Durban to meet it. When I arrived in Durban, the ship was there but the port was on strike. The machine was stranded for several weeks, and I had to come up with something. There was no sense in flying back to Sweden for a week just to fly back to Durban. I asked our local business partners whether there were any areas Volvo had not visited yet. They said that Zambia might be interesting; it has big markets. There were no communication or traffic connections to Zambia at that time, so I left my luggage in Durban. With me I took only a small briefcase packed with one shirt, some socks and underwear. I also wanted to see Livingstone's Victoria Falls, so I flew to Rhodesia, as close to them as possible. At the airport, I hopped on a bus that took me close to the Zambia border. I got off the bus in the middle of nowhere, and I walked several kilometers, crossing dozens of bridges. Thousands of Zambian refugees carrying their bundles and children passed me going the opposite direction. I was the only white person and the only one walking toward the border. I wondered where in the hell was I going?*

*At the border, the Rhodesian soldiers examined my papers and passport, and said that entry to Zambia was impossible. They asked me to wait. I sat at the border station for hours until I started getting bored. I thought there must be some way and decided to give it a new try. I slipped a ten dollar bill between my passport, and that did the trick: they immediately stamped my passport. Naturally, the bill had disappeared. The crossbar opened and the soldiers pointed me in the direction I should go. I walked for hours in the sweltering heat and dust. Late in the evening, I stumbled upon a hotel close to Victoria Falls. I got a room and food. Then it was time to go admire the falls.*

*The next problem was to get from the falls to Lusaka, where the Zambian Forestry Ministry was located. But there weren't any connections to Lusaka. I had heard that a Swede, Lars Adlers, worked there. During breakfast in the hotel, I met two Indian merchants who lived in South Africa. They had a car. I bummed a lift from them to Lusaka, where I managed to get Lars Adlers' phone number. The person who answered said he was on a very important forest visit. I received the coordinates of the location, and took off in search of a taxi, which then took me far from Lusaka to the middle of a forest. I started walking, and luckily found Mr. Adlers surrounded by local forestry professionals. I walked with them the whole day, examining the magnificent plantation, wearing a suit and with my briefcase in hand. It was as hot and dusty as hell. My fancy suit and briefcase were gradually covered with brick-red dust. I surely looked like anything but a convincing Swedish exporter.*

*At the end of the day, it really felt great when Lars Adlers invited me in his car and offered me a place for the night at his home. I am still grateful to him for that. When we arrived at his home, there was a surprise waiting. His wife had prepared for us two huge, ice-cold and refreshing gin tonics. I still remember the feel of dust washed away by gin. And then I had a shower, the first in a long time, and spent a really relaxing evening in Lars' home.*

*The next day — refreshed by the gin and the shower — I visited Lusaka's Forest Ministry. Lars' maid had brushed my suit into a presentable condition.*

*A few days later, I took the same route back to Durban, where the SM 868 was waiting on the dock, ready for work demonstrations.*

*The trip to Zambia was not in vain. After some time, we received from them a rather nice order for forest and earth-moving machines.*

*The best thing, however, was the gin tonic I got from Lars' wife. Nowadays, I drink this refreshing luxury only on special occasions. And whenever I'm holding the glass, I can taste the red dust in the back of my throat.*

HANS-ÅKE ELOFSON

In the mid-1960s, Hahn began to assemble a processing machine out of used parts to turn out short-wood pulp logs for Kimberly-Clark. When the company became aware of the machine he had built, it asked Hahn to build four more for its operations in Georgia. Because of his association with Cancar through his use of their skidders, he turned over fabrication of the machines to the Ontario equipment builder. When that arrangement fell through after a couple of years, Hahn decided to set up his own plant in Two Harbors, Minnesota, where his company has been building processors, known as Hahn Harvesters, ever since.

Hahn's basic machine was mounted on a rubber-tired trailer, which was easily towed to a landing or along a logging road. A knuckleboom loader fed a whole tree into the delimbing head, where hydraulically-operated jaws closed on it, and it was advanced by a hydraulic ram. The delimbing head pulled the log ahead a set distance, and the delimbed portion was cut off by a guillotine shear. Logs were collected in a cradle and loaded into piles or onto a truck. It was a simple, inexpensive, highly efficient roadside processor that found immediate success in many parts of the continent. In time, optional chain-type cutoff saws were added to produce sawlogs, and chain flail delimbers replaced the knives. The Hahn company later went on to produce slashers, delimbers and a single-grip harvesting head.

In the Pacific Northwest one other factor materialized during the late 1970s which was destined to have a larger impact on the logging industry than probably any single logging machine yet devised. From California to Alaska there arose a wave of environmental activism focused in opposition to the logging of old-growth coastal forests. Beginning with campaigns to preserve stands of redwoods in California, activists moved on to advocate the prohibition of logging on most of Moresby Island off the north coast of British Columbia. The Moresby Island dispute developed into a huge international issue, pitting a coalition of local natives and urban-based members of environmental groups against MacMillan Bloedel, Canada's largest forest company, and a few other timber-holders. By the time the dust settled the timber companies had lost, sparking similar clashes across the continent. The emergence of the environmental movement, combined in the early 1980s with the worse recession since the Great Depression, severely curtailed logging throughout North America and in many other parts of the world. As a result, the evolution of logging equipment slowed to a crawl for almost a decade.

In the early 1970s, a number of organizational changes in the Finnish forest equipment manufacturing business took place. Rauma-Repola, which had grown into a diversified company with assets in forest land, pulp and paper mills, sawmills, shipyards and engineering, acquired control of the machine manufacturing company, Lokomo, which was ready for a change. The diesel locomotive market had shrunk and was dominated by Lokomo's chief competitor, the state-owned Valmet. Immediately after the signing of the deal, a product development department was formed and a new factory built in Joensuu, where Rauma-Repola already had a sawmill, a wood works factory, land and an established reputation.

In 1972, Lokomo and Ösa collaborated on the development of a harvester built from an Ösa 750 processor installed on top of a Teli-Lokkeri. Lokomo had decided not to develop its own processor, but to move directly to production of a harvester, the Lokomo 960, which appeared in 1973. It felled, delimbed and bucked trees up to 60 centimeters in diameter and carried logs in bunks on the rear. In many ways the harvester reflected the fact that it had been designed by a committee. It was conceived through the joint efforts of experts from the National Board of Forestry's development department, Metsäteho, Enso, Tehdaspuu, the Forest Owners' Association and Hankkija. At the same time, several Canadians came to Finland to give their advice on the machine. There may have been too many contradictory opinions because the machine turned out to be somewhat neurotic. The target weight had been twenty tons, but the prototype weighed thirty tons. It had the power of a small village, and was capable of such a high speed that it was not always useful in Finland's small woodlot-based forest stands. Its numerous features meant that many things could malfunction. Great effort was expended on exporting it, but to Lokomo's disappointment, the 960 was no more of a success on the export market than at home.

The reorganized company dropped the Teli-Lokkeri name for its skidders, which were Finland's best-selling model, adopting the Lokomo name for its growing line of skidders and forwarders. By mid-decade it managed to penetrate the American market with the 933 forest tractor that

“

*I had a friend who became a lawyer that I stopped in to shoot the breeze with quite often. One day he says, “You know if anyone has a little money, there’s a lot of good deals go through here.”*

*“Well,” I says, “I got a little. If there’s a good one comes along, let me know.”*

*A while later he called and said, “You know, there’s kind of a wild Indian in here and claims he’s got an invention worth something. I don’t know if it is or not. Why don’t you come over and take a look at it?”*

*So I went over there and this guy had made this working model of a debarker and he had two or three little poplar poles. He’d spin ‘em right through there and peel the bark off. So I made a deal with him that if he’d patent it — ‘cause I didn’t know anything about a debarker — I’d get involved with it. Well, he was broke, he was going to lose his car the next day and he had all sorts of problems. So, anyway I give him some money that day and checked with the lawyers. He went and looked at it and said it looked good to get some claims on it. So that’s kind of where it started. It took maybe pretty near a year to build a machine. I think we sold the first machine in May to a company over here. They bought five machines for four thousand dollars apiece.*

*In the Lake states they debarked a lot of the wood with a drawshave. A guy with a drawshave, in the winter he could only debark about a cord a day. Maybe two cords in the summertime. Well this thing we built, two guys could run 15 or 18 cords a day. So it went over pretty good. That’s where we kind of got started, selling them all over the country.*

Norval Morey

had two axles at each end, giving it a load capacity of twenty tons, and a pulling power of twenty-six tons when it was equipped with a skidding grapple. The 933, known for a time as the world’s strongest forwarder, was an advanced machine for its time and was modified for various applications, including that of pulling cannons for the military. A tracked version was built for operating on soft ground and in swamps. It was the favorite machine of many operators because it was almost impossible to roll and it could be driven anywhere.

A partnership between Finnish and Swedish interests developed at about the same time, just after the cooperative arrangement between Lauri Marttiini and Valmet ended. The Finnish designer from Rovaniemi and Volvo joined forces to build a new forwarder, the Volmari. The front end from a Volvo was mated with a trailer developed by Marttiini, but it was not a success. It had a rigid frame and did not maneuver well in the snow, so only a dozen machines were sold before production was stopped.

In the late 1960s, a Finnish logging contractor Einari Vidgrén became fed up with the breakage problems of available machines and built his own forwarder in his brother-in-law’s machine shop in a small village near the northern city of Iisalmi. He put it to work in the forest where it proved to be more durable than other models then in use. The reputation of the machine spread, and other loggers began requesting one. Vidgrén set up a plant, and in 1971 the first Ponsse PAZ drove out the factory door. The letters PAZ came from the Perkins engine, the Allison transmission and the ZF planetary axles. In spite of the popularity of the machine, Vidgrén did not become a full-time manufacturer, having many other irons in the fire. He drove race cars, established a race-horse stable, ran a logging company, sat on the county council and occasionally filled in as a machine operator. PAZs were built at the rate of only five per year; by 1980, sixty of them had been sold.

In 1972, Sakari Pinomäki, whose Pika 50 processor had been so well received, added a new feature to his machine and produced the Pika 75, which was capable of controlled felling. It had a 5.7 meter knuckleboom equipped with a chain saw for felling and bucking, and a separate delimber. Pinomäki called his machine a harvester because it converted a standing tree into a pile of logs. Although the Pika 75 was not a commercial success, the concept was aimed in the right direction.

Pinomäki also worked with Valmet to build the 880S feller-skidder. In the early 1970s, Valmet had two lines, one of big forest machines and another, Valtras and Jehus, built on farm tractor bases. By mid-decade, the Valtra line was phased out and the Jehu, built at the Rovaniemi plant, had acquired an off-road transmission and 115-horsepower turbocharged engine. This machine complemented the company’s larger machines, which included the 882K used as the base machine for a forwarder, skidder and feller-buncher.

In addition to the big factories, small manufacturers were adding new machines to the market. Lapland still needed machines capable of working in deep snow. Northerners were of the opinion that southern manufacturers did not understand Arctic conditions, so they resorted to their own solutions. At the Rovaniemi Machine Shop, Lauri Marttiini modified farm tractors by converting them to frame steering and adding half-tracks. In 1974, Muisto Laine set out to develop an inexpensive, Bobcat-based feller-buncher for harvesting pulpwood. The first version was a failure, so he added some more components and ended up with a harvesting machine, the Makeri, which could fall, delimb and buck trees into shortwood pulp logs. It became a popular machine, and later a forwarder of a comparable size was developed.

In the mid-1970s, Finnish machine companies began to pull ahead of their Swedish counterparts due to different economic policies pursued by the governments of the two countries. Sweden, which had enjoyed a lead in export markets since the turn of the century, began to lose ground because of inflation, while Finland adhered to a tight economic policy which improved its export potential.

Other factors limited exports of Nordic forest machinery to European customers. The new Nordic machines were generally too large for the small stands found in most European countries, and not well suited to the primary form of logging, which consisted of small-scale thinnings. As a consequence, most of the work was performed with domestically-produced machines. Poland and Czechoslovakia expressed an interest in purchasing Nordic machines, but their lack of hard currency prevented sales. Cable logging systems were built in Italy, Spain, France, Switzerland, Norway and Czechoslovakia. The Soviets mounted Lutj and SM-2 delimbing machines on TDT-55 tractors, and were developing processors.

In Japan, developments followed a different course. With approximately the same forest area and annual growth rates as Finland, and a strong silvicultural tradition dating back to the fifteenth century, Japan's post-war evolution took a dissimilar turn. Following World War Two, the country's timber-producing industry went into a state of decline as imports of cheap logs from North America and Southeast Asian nations grew. In spite of a serious decline in the number of available forest workers, the country was slow to mechanize, especially in contrast to developments in other industrial sectors. A domestic chain saw appeared in 1947, but in the 1950s was replaced with lighter US models. During the 1950s, forest railways were replaced with trucks, and late in the decade Iwafuji developed a crawler tractor for forest use. Beginning in the 1960s, skidders came into use on the lower, moderate slopes of the country's mountainous forests, while mobile cable yarders were employed exensively on the steep upper slopes. In 1967, Iwafuji introduced a wheeled tractor that was used as a skidder. A widespread mechanization process occurred during the 1970s and 1980s. By 1985 there were about 20,000 cable yarders in use, along with 4,000 skidders. In recent years about 16,000 forwarders have been added.

In Sweden, much of the focus during the late 1970s was on the development of larger and ever-more-sophisticated forest tractors as base machines for processors, fellers, feller-bunchers, skidders, forwarders, scarifiers and planters, many of which remained popular until the mid-1980s. Kockums' 850 forwarder, a 14-ton model, was typical of this period, as was the Volvo 980 and Ösa's 260, the first Swedish forwarder with a hydrostatic power transmission. More than 850 of the latter were sold during a ten-year period, primarily because of its versatility and the popularity of its operator-friendly cab.

The environmental concerns which had begun to emerge in North America at this time were also appearing in Scandinavia, producing a need for thinning machines to provide an alternative harvesting method to clearcutting. The recently-evolved big forest tractors were unsuited to this task, and a new generation of smaller machines emerged. One of the earliest was the Forest Ant, a machine the size of a large garden tractor designed as a carrier for a feller or a feller-skidder in thinning applications. The operator walked alongside and operated controls attached to the tractor with a cable. A prototype was built in 1975, but the Ant never made it into serial production. Three years later, Bruun Systems brought out the eight-ton Mini-Bruunett 578F forwarder. It was dismissed as a toy by proponents of big tractors, and criticized for being too large by those looking for a compact thinning machine. At the same time, Doroverken introduced the Little Skotten 747. Both machines were small and maneuverable with eight wheels and high flotation tires for low impact on soft ground.

For a few brief years, strong forest product markets raised fears of a potential Swedish timber shortage, which encouraged the development of machines to improve wood utilization and reforestation. The Skogsarbeten initiated a project to develop techniques for the total utilization of trees but it only managed to develop one device, the short-lived Mackmyran stump extractor.

Perhaps the most significant Nordic development of the late 1970s was the first Scandinavian single-grip harvesting head. One of the first of these new devices was developed by a small Swedish firm, Rosenberg and Kruuse of Halmstad, which built small tractors for Volvo. The single-grip harvester was conceived as an alternative to the two-grip harvesting system then in use, in which a boom-mounted felling head cut the tree and fed it into a processing unit mounted on the machine's chassis. In 1978, the RK Skogsjan 450 processor appeared, utilizing the new harvesting head. Mounted on a forwarder carrier, it consisted of a boom and a harvesting head with a pair of hydraulic clamps which held the tree while it was severed with a chain saw. A set of rotating steel wheels fed the tree through a set of delimbing knives, and the saw bucked it to length. After producing about ten heads, Rosenberg and Kruuse sold rights to the new harvesting head to Volvo. At about the same time they developed the Skogsjan, a Finnish version was built by Sakari Mononen, called the Finko.

> “
> *Sometimes the financing was taken from the contracting company. We didn't have any kind of marketing. We thought that a good machine will sell itself. And whenever we had a machine ready, there was a buyer for it. We are little like Rottne in Sweden. Ponsse also originated from contractors. We are folksy. We are located here in the middle of a forest, and we don't have to brag about being big. We were going to build a new office building, but then thought that we won't change this image, even if we are a little cramped here.*
>
> Einari Vidgrén

As a trend toward higher quality timber production evolved, attention focused on the damage done to logs by harvester heads, particularly by the feed rollers. The early rollers utilized steel spikes to grip the tree, but these left deep marks in the most valuable, outside part of the log. One solution was to use rubber wheels for feed rollers.

The rapid evolution of the single-grip harvesting head was partially due to the development of small tractors suitable for thinning. The second factor involved the introduction of computers in processor heads. Before this development, log measurement to facilitate automated bucking had depended on the use of electric relays and sensitive microcircuits which were cumbersome and prone to breakdowns caused by vibration. Their replacement with microprocessors encouraged the rapid evolution of automated measuring and bucking technology. In 1978, Volvo's new 900 harvester featured an automated measuring system using computer technology, and the following year Bruun Systems developed the Bruunomatic automated processing head. Mounted on its Mini-Bruunett, it was the first small-scale forest machine able to thin forests in an economic manner without damaging the remaining trees or the forest floor. The combining of the various new systems in the single-grip harvester soon led to the widespread adoption of mechanized selective harvesting. In 1975, only five percent of thinning harvests in Sweden were fully mechanized; by 1992 the proportion had increased to 68 percent, and the volume of timber obtained by thinning had almost doubled.

As was the case in North America, the 1970s had seen a flowering of Nordic felling, delimbing and bucking machines, combined in a variety of processing and harvesting machines. By the end of the decade, forest mechanization was advancing rapidly. In Sweden, for instance, 6,000 forwarders had been sold during the previous ten years, seventy percent of them produced by Volvo and Kockums. In Finland, Lokomo and Valmet dominated the smaller local market. Many people believed this mechanization process would continue unabated into the 1980s. Others, who were aware of broader economic factors, realized that major structural changes in the forest equipment business were required. The global recession which subsequently occurred hastened that restructuring.

A Cypress grapple yarder working on the British Columbia coast in the early 1980s. Lights are mounted on the top of the boom for night logging.

A Mini-Alp cable yarder mounted on the rear of a Timberjack skidder. This combination was developed to thin Douglas fir stands in British Columbia in 1980.

The Ecologger cable yarding system was a small, portable steel spar built in British Columbia during the 1970s and 1980s, and used around the world.

The Skylead Logging system consists of a portable spar mounted on a variety of carriers. It employs a skyline and, as shown here working in the British Columbia interior, powered carriages. It has been built in British Columbia since the mid-1980s and is used globally.

An amphibious Weasel piles logs at a landing in a Columbian rainforest in the 1970s.

A stroke delimber working in Quebec in 1988. Developed in Sweden in the 1970s, this type of delimbing machine quickly became popular in parts of North America.

This crawler-mounted steel spar and knuckleboom loader is yarding logs up a steep hill and loading them on a truck near Yaak River, Montana in the 1970s.

The Currie Cost Cutter, bucking a load of logs into pulpwood lengths in the southern US in the 1970s, provided an easy means of reducing tree-length logs into short pulp logs. Known as bundle buckers, these machines used a giant chain saw mounted on a gillotine to cut through the entire load.

This Hydro-Ax chain flail delimber evolved from a prototype developed by Gordon Frankin in Manitoba, Canada, in 1972. It was driven over trees spread on the ground, but did an incomplete job.

The limbing gate, widely used to on southern US pine, is one of the simplest and most effective delimbing devices. A grapple skidder backs a turn of logs through the grid, breaking off the limbs.

The Westvaco feller buncher was a prototype built by Joe Blonski in 1966. A modified version of the Vit's feller skidder, it sheared a tree, which fell over the machine and hung on the overhead toothed bar. When a load was accumulated, it was dropped on the ground.

In Scandinavia during the 1970s, clear cut forests were plowed, harrowed and scarified for reforestation. Pictured is an Ösa 250 pulling a Donaren scarifier.

The small Swedish Skogs-Basen skidder was designed to replace a horse in the woods. The idea was good, but performance figures were low.

The idea of a single-grip harvesting head appeared simultaneously in Sweden and Finland in 1977. This Finko head began as a delimbing and bucking head, and later a falling saw was added, launching an era of single-grip harvesters in Finland.

The Swedish Skogsjan processor went through a similiar evolution as the Finko, and was eventually bought by Volvo.

In the early 1980s big processors — such as the Kockums in the picture — gave way to the new single-grip harvesters.

The Combi processor-forwarder was developed in Sweden by Husqvarna in the 1970s. The basic machine was a Rottne. Trees were still felled by chain saws.

A John Deere two-grip feller and delimber. A falling head mounted on the knuckleboom felled a tree and fed it into the delimber on the rear of the carrier.

Ösa's stump extractor. It efficiently removed stumps, but they were too contaminated with rocks and dirt to use for chips.

A tropical hardwood tree is felled with a chainsaw and skidded with a Timberjack skidder at Jari, Brazil, in the 1970s.

The Ösa 260 forwarder was a success during the 1970s in the mid-sized category and was very popular as a base machine.

Rottne's Blondi has been called the most beautiful forest tractor of the 1970s.

Lars Bruun's Mini Bruunett appeared in the late 1970s. The small forwarder had 8 wheels, weighed 8 tons, was terrain friendly and very agile.

The Mini Bruunett 578 processor was a great success into the 1980s.

The new-generation machine combination in the mid 1980s, the Ponsse harvester and Norcar's forwarder.

# The Modern Era

The recession which began in the late 1970s hit the global forest industry particularly hard. It was accompanied by soaring interest rates which substantially reduced new construction, and the logging industry followed the lumber industry into a prolonged slump. Although the economic downturn seriously reduced the sales of logging equipment, along with spending on research and development, the evolution of timber harvesting machines continued, if only in the minds and on the drawing boards of designers.

In Sweden, Skogsarbeten suggested there was too much diversity in forest machines, causing high repair and operator-training expenses. It recommended that fewer, standardized machines be developed. Elsewhere, a number of evolutionary processes, begun in the previous decade, continued through the early 1980s, then advanced rapidly as the industry recovered in mid-decade.

In North America, one of these processes concerned the evolution of the feller buncher. This machine was the product of several other developments. The move to full-tree grapple skidding created the need for a machine to fall trees and stack them in turn-sized piles, oriented for quick skidding to the road. The abandonment of harvesting machines, such as the Busch Combine, created the need for a new type of mechanized falling machine, and encouraged the technical development of various falling and whole-tree handling heads. In turn, the development of new cutting heads gave further impetus to the refinement of the feller buncher.

The enthusiastic response to the first Drott feller buncher encouraged other manufacturers. One of the first to take up the challenge was Koehring Waterous, which had some experience in the field with its earlier harvesters. In 1978, under chief engineer John Kurelek, the company first attempted to adapt an excavator, a course also followed by Caterpillar and John Deere at about the same time. In spite of a beefed-up cab and engine covers, added track protection, a new boom and a shear-type felling head, this machine was not a success. In 1981, Koehring completed its first purpose-built tracked feller buncher, the Model 620, which overnight became the market leader and established the benchmark for tracked falling machines. In 1988, after Koehring was purchased by Timberjack, two new models were introduced, the 618 and 628. A year later these machines were sold under the Timberjack name.

On a different front, a rubber-tired, drive-to-tree feller buncher was introduced by Hydro-Ax in 1976. It was designed by Len Vizina who started working in the woods when he was twelve, pulling a crosscut saw with his uncles in Michigan. In the mid-1950s he graduated to one end of the first two-man Mall chain saws, a job that convinced him to get out of the operating end of the timber falling business. He went to work for Pettibone in 1958 and over the next fourteen years worked on the development of skidders, slashers, grapples and other equipment. In 1972, he left to work with a tree-clearing business to develop a hydrostatically driven tractor on which to mount a brush mower. With his background in forestry, Vizina soon figured out ways to attach logging equipment to the tractor, and a new concept in feller bunchers was born. The Hydro-Ax feller buncher was an operator-friendly machine, with quick responses and the ability to maneuver easily in the woods. The key to this capability was hydrostatic drive, which dispenses with driveshafts and axles and utilizes hydraulic motors in each wheel. The speed of hydrostatically driven machines can be precisely controlled, but early attempts to develop them encountered major problems with overheating of the hydraulics. Vizina's machine was perhaps the first to overcome this problem. It was particularly well suited to plantation forests in the southern U.S., where ground conditions permit the machine to readily approach each tree. Hydro-Ax quickly took the lead in supplying the growing market for this type of feller buncher, and received an additional boost in popularity when shears were replaced with high-speed disk saws. By the time the forest industry got back on its feet in the mid-1980s, several established manufacturers, including John Deere and Franklin, had added hydrostatically-driven feller bunchers to their lines.

> *We saw a lot of people start out with just the directional felling shear, like the old I.B. White shear. We used to put a lot of them on skidders. They'd just fell a tree, and even though a lot of them were still using cable skidders that was a big advantage because everything was felled in the same direction. So they could just back up to the stumps and they didn't have to keep turning the trees around. I think that was the original concept of the shear. When the feller buncher came along, the grapple skidder was just a natural addition to it. Then I think it was like the evolution of any piece of equipment, the need was there and the ingenuity took over and created the product.*
>
> Harold Estes

Another type of mechanized feller buncher, a three-wheeled machine, also made an appearance in North America in the late 1970s. Several years earlier, a South African mechanic, Irvin Bell, had designed a version of this machine to load sugar cane. When the South African pulp and paper industry was established, Bell's loader evolved into a more versatile machine used for loading pulp wood. The next step was to hang a directional falling saw on it, which was when it appeared in North America. By this time three of Bell's sons had more or less taken over management of his company, and they made a deal with Norval Morey for Morbark to build a model of the Bell feller buncher for the North American market. When the Morbark-Bell machine did not catch on, the agreement expired. The Bells then set up a small assembly plant in Savannah, Georgia, and worked with Ev Stephenson at Union Camp to develop a modified version, which was ready for market in 1983. Because of its low cost and simple mechanics, this machine eventually became very popular with flat-land loggers.

The Bell machine was not the first three-wheeler to come onto the market — Timberline Equipment had sold one for years — but it was the first to succeed. What it did, and did very well, was essentially what a man with a chain saw could do. It could directionally fall a tree and, within limits, run with it to control where it ended up, much like a child balances a baseball bat in the palm of his hand, running to keep it upright. The Bell was a mechanically simple machine, less expensive to build and operate than most feller bunchers on the market. Its success prompted several other U.S. companies — Franklin, Hydro-Ax, Gafner and CTR — to introduce tri-wheeled versions. Morbark subsequently came out with a three-wheeled tractor, the Wolverine, fitted with a variety of attachments, including a range of feller-buncher heads. Like the Bell, its use was limited to fairly easy ground conditions, primarily in the American South.

At about the same time Len Vizina was building the first Hydro-Ax feller buncher and Koehring was producing its first tracked machine, another former chain saw operator was developing quite a different falling machine. Pat Crawford began his career as a third-generation logger in northern Wisconsin. After a couple of years in Oregon working as a faller, he returned to Wisconsin and set up his own company to log on national forest land. He bought one of Leo Heikkinen's first hydraulic loaders and caught the equipment-building bug while working with Heikkinen on an improved version.
He also got one of Erv Drott's first feller bunchers, and the two became close friends.

The problem Crawford found with the Drott machine, common to all excavators used for this task, was that it was not designed to do selective cutting, a capability he needed on the national forest and Indian reserve lands he was logging. These machines overhung their tracks, leading to unacceptable damage to the trees left to grow. He set about designing a different machine and built his first one in 1978, in a machine shop near Shawano, Wisconsin. He had no intention of becoming a manufacturer, and tried to convince Caterpillar, for whom he was doing some testing, to take over the design. When the company declined, Crawford hocked everything he owned to raise money to set up his own plant, Timbco Hydraulics. He came out with his first feller buncher in 1980.

Crawford's feller buncher utilized two unique concepts. First, he moved the pivot point of the main boom and the anchor point of the main lift cylinder behind the center of the tracked carrier to eliminate the need for a counterweight. That, and the tight configuration of his boom, meant the entire machine rotated within the width of the tracks, enabling it to work freely in standing timber, selectively falling and bunching trees.

Crawford's second innovation was a levelling cab and boom. His first Timbco levelled two ways, on a parallel axis, and his second model, which appeared in 1982, levelled four ways, on both the parallel and transverse axis. This capability allowed his machine to operate on steeper slopes, up to fifty percent or more — beyond the capability of any other feller buncher on the market. Crawford demonstrated the machine at a show in Spokane, where it caused a sensation. Unable to keep up with the demand, he entered into a co-manufacturing agreement with Timberjack, which was terminated in 1987. A few years later his small plant in Shawano was turning out almost 300 machines a year, and exporting them all over the world.

Then, late in the decade, another new entrant came on the scene. In 1987, when Timberjack acquired the Koehring Waterous company, a group of Koehring Waterous engineers — including John Kurelek, one of the key engineers on several major Koehring projects — found themselves out of work. They approached MacDonald Steel, which manufactured components for several equipment builders in southern Ontario, and convinced MacDonald to set up Tigercat Industries to build logging equipment, beginning with a feller buncher. They came out with their first, tracked model of the Tigercat in 1992, and a second, rubber-tired model a couple of years later. The success of these machines encouraged the company to fill out its product line with others, including a hydrostatically driven rubber-tired skidder, which are still on the drawing board.

"

*One of the most interesting developments was the satellite chip mill, when it came out and began to accelerate in the '80s. What was going on there was that the pulp and paper companies were beginning to contract directly with the loggers and independent contractors, leaving out the middleman or broker who was the wood dealer in the past. When we had primitive guys with chain saws who weren't well schooled in business management, we needed those guys in the middle.*

*As the independent logger became a better financier, a better machine manager, a better business manager, then we could work directly with him. The wood dealer — that prominent guy out there in the small town with all the banking connections and the management ability — went looking for other things to do. One of the things he found out he could do was to go into the independent processing business, set up a satellite facility for debarking and chipping wood. Then he could buy the wood standing, still have his guys log it, because they work for him now, and debark it, ship it, and sell the finished product to the companies, the large companies who found a number of reasons to promote this. Green-field mill sites are still difficult to develop for environmental and other cost reasons. We've never fully optimized the benefits of scale of most of our mills. We could add on capacity at the mill rather economically and we found that we could export our debarking and chipping business off to him and expand the mill over into the woodyard and make more paper and make more money and get out of the capital investment, get out of the management, get out of the people business. So one of the major things in the South was the explosion of those guys, over a hundred of them going in over the last 15 years.*

Ev Stephenson

One of the major changes in feller-buncher technology during this period was the introduction of a new type of cutting head. The first feller bunchers used either a hydraulic shear or one of the chain saw heads first utilized by Rudy Vit on his harvester. The shear, which worked well in the South, was not suited to northern forests because it produced an unacceptable amount of shatter when cutting frozen wood. In the late 1970s, both Hydro-Ax and Drott developed rotary cutting heads for their feller bunchers, but they were not successful either. These ideas were picked up and subjected to intense research and development by Koehring Waterous, under engineer John Kurelek, and the Forest Engineering Research Institute of Canada (FERIC). The result was a patent issued to Koehring Waterous in Ontario and Harricana in Quebec, both of which began manufacturing the saws in the mid-1980s. The large circular blades on these saws are edged with carbide teeth; driven by powerful hydraulic motors, they can slice through a tree in a second or two. Initially designed to cut frozen northern wood, they gradually took over as the most popular type of mechanized falling head for feller bunchers. In 1988, Georgia Pacific in the US stopped accepting shear-cut logs, giving disk saws a big boost in the South.

The use of feller bunchers spread even to the West Coast in the 1980s as the harvesting of smaller, second-growth timber increased. They were often paired with grapple yarders to move bunched logs to roadside. In the mid-1980s, Alberni Shipyard on Vancouver Island built the local Giesbrecht logging family an oversized feller buncher from a beefed-up Timbco which could cut and handle a 75-centimeter tree, forty meters tall — the largest feller buncher made to that time. A few years later, Madill, the long-time manufacturer of steel spars and grapple yarders on Vancouver Island, came out with a large feller buncher capable of handling West Coast timber.

Another field of development, which continued through the recession and ended with production models appearing in the late 1980s or early 1990s, was a range of processing machines. Advances in this area were pushed by a number of factors. With the continued decline in the operational capabilities of US railways, there was a general movement toward trucking as the favored long-haul method of moving logs from the landing to the mill. At the same time, increased regulation of the trucking industry was making it more expensive to haul logs, particularly tree-length logs on public highways. This encouraged a greater degree of processing in the woods or, in some cases, at satellite wood-processing yards.

Throughout this period, with increased use of feller bunchers and grapple skidders, there was a consequent increase in the use of roadside delimbers, especially stroke delimbers. In 1982, a new concept appeared when Jean-Paul Tanguay, a Quebec machine-shop operator and inventor, produced a knuckleboom delimber. Tanguay had spent several years developing and building slashers, loaders, feller bunchers and other logging tools in his small shop. His delimber dispensed with the stick or rail which was a feature of most machines of this type, and utilized a clamp, which grasped the butt of the tree, and a floating delimbing head mounted on an articulated boom. The head followed crooks and sweeps in the tree much more easily than a stroke delimber, giving it superiority in some types of timber.

A simple and less costly delimber appeared in the early 1980s, the pull-through delimber. It consisted of a set of delimbing knives mounted on a frame which was set up at the landing. During the loading process, a knuckleboom

> “
> *In 1991, the Scottish contractor James V. Lapsley ordered four forest machines from us to be sent to Russia, east of the Urals. They had to be delivered quickly and safely. Rail was not acceptable because parts always disappeared from the machines during train transports. But Lapsley got things rolling. He chartered the world's biggest cargo plane, an Antonov, to transport the machines. The Tampere airport is too small for the Antonov, so we took the forest machines to Helsinki. It was a great moment seeing the Antonov take off into the skies with four forest machines in its hold. The shipping agent's men wondered how we came up with that idea. The incident is noted in Finnish aviation history and it received publicity in the media, but it also had a down side. The machines with their sophisticated technology were taken to a military airport. Even then, any part that could be detached was stolen.*
>
> Pertti Huusko

loader was used to pull the tree through the delimber, then load it onto a truck. CTR Manufacturing produced one of the first versions of this delimber and has since been joined by several other North American companies, including Bell and Morbark. They have been particularly popular in the southern U.S.

To some extent the acceptance of mobile, woods-based processing machines was made possible because of the increased multi-functional capability of other machines, particularly loaders. Knuckleboom loaders had evolved so that they could feed trees into slashers, debarkers, and delimbers, as well as load logs onto trucks.

The very concept of multi-functionalism had been on the decline for several years in North America after the unsuccessful experience with harvesters twenty years earlier. Now, one of the main weaknesses of these earlier machines, their lack of reliability, had been overcome. Machines such as the Busch Combine had spent thirty to forty percent of their time in an inoperable condition. By the 1980s, with new technologies for controlling machine processes available, the operation of various machines became easier and more reliable. For example, the original knuckleboom machines utilized a combination of foot pedals and levers to control the machines' movements. As well as requiring a great deal of skill on the part of the operator, these controls were undependable. The installation of joy-stick controls dramatically improved their reliability and ease of operation.

Today, the knuckleboom loader and its operator have assumed a far more complex role than they played a decade ago. The operator can take raw logs — whole trees, with limbs and tops — and feed them into a variety of processors to produce whatever product is desired. Depending on what the market demands, he can process various species into pulp or sawlogs, delimb or debark them, buck them to whatever length is wanted, or feed them into a chipper. The loader operator has become probably the most critical member of the logging team.

One of the latest tools in his arsenal is a portable whole log chipper. Until the late 1980s and early 1990s, chipping in the woods was rare. The production of chips to feed pulp mill digesters was a function that conventionally was performed in the mills by large, whole-log chipping blades. Bringing this technology into the woods was seen as difficult, if not impossible. The other obstacle to the production of pulp-grade chips in the woods was the lack of debarking methods to prepare logs for the production of clean chips, particularly in the case of softwoods.

In the 1980s several companies, including Morbark and Peterson Pacific, introduced the first generation of portable chippers capable of producing pulp-quality chips. At about $500,000 US dollars each, they are among the most expensive machines in the woods today. The operation of most of them is similar. A grapple skidder drops a turn of logs, which can include saplings as small as four to six centimeters in diameter, within reach of the loader. Several stems at a time are fed into the processor, which first runs them through a chain-flail, debarking the main stem and branches down to about a two-centimeter diameter. The bark, needles, leaves, limb tips and tops are ejected from the machine and used for fuel, mulch or some other product, or are hauled back into the forest and spread to provide nutrients for the growing crop. As the tree is debarked, it is fed into the chipping blade. Chips are directed into a waiting trailer at the rate of 500 tons or more a day. Portable chippers increase the amount of fiber that can be utilized from a site by as much as twenty-five percent. They are finding a ready acceptance for several reasons: It is

“

*FMG's Lars-Göran Kämpe was in Canada to train the locals in the secrets of the Lokomo 990 harvester. The 990 had the Finnish-developed Lokomatic measuring automation, which was still somewhat unfamiliar to the Swedish Lars. The size of a giant, Lars looked assertive when climbing into the driver's seat. He fired up the engine with confidence and then turned on the measuring machine. The date appeared on the display first, and then the text: "Do you want to continue?"*

*Lars became puzzled and didn't notice the text: "Yes or no."*

*Some other FMG worker offered Lars advice: "Say yes."*

*Lars bent down right next to the display and whispered in a friendly voice: "Yes!"*

Hans-Åke Elofson

“

*In the late '70s I met Russell Stevenson, a South African who came over here representing the Bell brothers. Their father was a mechanic, working for one of the wealthy plantations there in South Africa, who began to innovate in the middle of his career. Then went out on his own, near retirement age. There were three brothers who grew up and each got a reasonable technical and business education and they rather quickly turned it into a far more professional company. They started off in sugar cane equipment. The little three-wheel tractor was designed and conceived by the elder Bell, Irvin. The sugar cane worker there was accustomed to walking with his feet and doing things with his hands and Irvin could see that moving forward by pressing a foot down on a treadle and moving back by pressing a heel down would be a rational thing as a control for untrained people. You weren't going to teach them how to run it, you were going to turn them loose with it and let them kind of figure out how to run the thing, what did what.*

*So it worked. They built primitive little three wheelers with single- and two-cylinder gas and diesel engines out of pipe tubing and hung large grapples on them. The sugar cane cutters were cutting sugar cane and they would cut in a circle and tilt it together in a tepee-like shape around the swathe they cut. These machines could come up, pick it up, trundle it over, and put it on a cart. A loader of sugar cane, batch loader, I guess was the original thing. And then they made it more of a materials handler, hanging everything on it they could. Then they went from that to loading short wood when the pulp industry started down there. So they put a smaller grapple on it and they were loading short wood. Then the concept of hanging the directional felling saw on it — the Scandinavian style directional felling saw — just occurred as another tool to hang on it.*

*When I went there in '86, they had 16-inch felling heads mounted on some of those tractors clear-felling pine plantations. I didn't know at the time how big a tree you could handle, and whether or not you could control pine saw timber particularly, or hardwood saw timber, and exert enough directional felling force on it to directional fell it without splitting the tree before the saw completed its cut. But the answer was yes, with a little bit of practice you could certainly do that.*

*Where the energy is lost in our current process is using huge, expensive, heavy machines that have to be big enough to cut and carry a tree on all kinds of terrain. This little machine does nothing more than what a man with a chain saw could do, pretty much — directional fell it. Although, while it's in the process of falling you can run with it, kind of like balancing the baseball bat in your hand, move around and get it where you want it. It worked quite well.*

*I ended up kind of working with Bell. I went to South Africa in 1986 and described to them what I saw as the potential of it. In that period of time the only saws that we had were very expensive, clumsy pieces of equipment, high production, high maintenance. The Bells had tried something that was pretty unique, hanging a Scandinavian style saw on one of their little three-wheelers. Paul Bell had put one of those together here while he was still at Morbark, but never even took it off the trailer for testing. It looked so ugly and clumsy, nobody wanted to see it, so he never even unloaded it.*

*In 1986, I found either four or six of those. I did some quick analysis and found that if those would work and if you could Americanize the concept — the combination of the three-wheel machine and the cutting head — make it suitable here for the conditions and people, that it should be far more productive and cost effective than anything we had on the drawing board.*

*So Paul Bell moved into my spare bedroom and spent a year in prototype to product. We ran it on our company crews. Brought the second generation of it out. He put together a facility here in Savannah to import, do the final assembly, and distribute, and for a couple of years had the principal part of the market in that area.*

Ev Stephenson

often cheaper to truck chips than logs; a truck box can be completely filled with chips, while a truck load of logs contains a lot of empty space. As well, pulp mills are finding it increasingly difficult to dispose of residues from the chipping process. Some companies, such as Union Camp, have calculated that the cost of spreading residues back in the forest, when logs are chipped in the woods, is equal to the economic value of these residues as fertilizer, so it is increasingly desirable to leave the unwanted residues in the forest from the outset.

In the early 1980s, alarming news for forest owners came out of central Europe. Sulphur emissions from fossil fuel consumption were increasing atmospheric acid levels and killing trees. This news touched off widespread concern about forest conditions generally, particularly in Scandinavia. In Finland, clear-cutting became a focus of public criticism, and forest land owners began to object to any type of logging activity which they felt damaged their land. The result was a growing interest in smaller, lighter machines.

Once started, the public concern with forest practices was not easily put to rest. Throughout Europe and North America agitation against standard timber harvesting practices spread, culminating in consumer demand for forest products certified to have come from forests managed according to a specified set of standards. This development also increased the interest in more site-sensitive machines.

One of the consequences of the economic upheavals that shook the forest industry during the 1970s and 1980s was a widespread corporate restructuring of North American and European equipment companies. Mid-sized manufacturers were particularly stressed, precipitating a flurry of courtships between firms seeking partners. The result, from the mid-1980s on, was a series of mergers and acquisitions that substantially altered the profile of the industry, leading in some cases to compatible matches of equipment types and in other cases to the acquisition and development of entirely new systems by some of the larger players.

The blending of Prentice loaders and Hydro-Ax feller bunchers was typical of the type of operation which emerged out of this restructuring. Prentice, which began as the Heikkinen Machine Company in 1945, was acquired by Omark Industries, an Oregon company, in 1967. In 1980, faced with a depressed market and the need for capital to develop its popular new feller buncher, Hydro-Ax was sold to Omark, and various functions of the two companies — marketing and product development — were combined. Then, in 1985, Omark was purchased by Blount Inc., which was founded as a construction company in 1946 by Wynton Blount, and later converted into a diversified manufacturing company. The Omark purchase was its first foray into building logging equipment. In 1990, manufacturing rights to Tanguay's line of timber harvesting and handling equipment were acquired, along with a marketing agreement with Norcar, of Finland, for cut-to-length harvesting equipment. Three years later, Blount acquired CTR Manufacturing with its line of pull-through delimbers and slashers. Today, all these companies continue to produce equipment using their original names, under the Blount umbrella.

A similar restructuring process was occurring in Europe. Ponsse, which had produced PAZ forwarders through the 1970s, largely for a local Finnish market, went into an expansion mode, bringing out a series of tractors with radically new aluminum frames in the early 1980s. These tractors, the S-15 and S-20, were enormously successful and in 1986 Ponsse came out with a falling head, the 520, which performed poorly. The following year the company purchased an interest in Kajaanin Automatiikka, with which it developed its own log measuring technology. Then, a year later, Ponsse owner Einari Vidgrén sold his family business to Norcar. In Sweden the Rottne company, comparable to Ponsse in its independent stance and local base, avoided the same fate as its Finnish counterpart by expanding. In 1988, it bought the former Stensele Mekaniska Verkstad factories in northern Sweden and also began producing its own automated measuring equipment. On the other hand, Hemek, another Swedish company which evolved from a local machine shop to produce forwarders and a single-grip harvester, was taken over by Hasab in 1989.

*"When the walking forest machine's test model was ready for tests, we wanted to do the tests secretly. We found a remote back forest that fit our purposes. But to our dismay, in the middle of the forest was a cottage that was inhabited by an old woman, a hermit who had received life-long rights to live there. We thought hard about how to explain to her the kind of thing she might come across in the woods, so that she wouldn't get scared or rush to tell everyone. We came up with all kinds of elaborate stories, but then we decided to just tell her what the machine was, and ask her not to say a word to anyone about it. We thought that if the old woman told someone in the village about a walking machine in the forest, no one would believe her anyway.*

ERKKI KARE

In North America, the parentage of the pioneering Tree Farmer skidder line underwent a major change in 1990 when Hawker Siddeley sold its skidder manufacturing division to Franklin Equipment. Tree Farmer had been established in 1960 at Enumclaw, Washington, by Dwight Garrett. The following year, Can Car, a Hawker Siddeley subsidiary in Fort William, Ontario, purchased the right to build Garrett's skidder everywhere except the northwestern US. These machines were sold as the Can Car Tree Farmer. By the late 1980s, the Garrett company was gone and Can Car had ceased building Tree Farmers. When the Drake family took over the Tree Farmer operation and its manufacturing plant in Talladega, Alabama, they merged the operation with their Franklin line of equipment, retaining the Tree Farmer name.

The economic downturn also produced some marriages of convenience among companies in the same or similar line of business. For example, two Quebec manufacturers of processing machinery, Denis and Harricana, merged in 1991 to form Denharco.

The most interesting mergers of this period were probably those between European and North American firms, creating a number of global manufacturers of forest equipment. One of these involved Timberjack in Ontario and a group of recently-joined Finnish and Swedish companies. Almost a decade after it was founded as Timberland Machines in 1947, the company introduced its Timberskidder, which propelled it into the big leagues of the equipment business. Needing money for expansion, the founders sold control to the Baltimore-based Ellicott Corporation, a manufacturer of hydraulic machinery, in 1960. The rapid success of the Timberjack articulated skidder led to the creation in 1964 of a subsidiary company, Timberjack Machines, to manufacture logging vehicles. In 1967, this company was purchased by the Eaton Corporation of Cleveland, which retained control for seventeen often-turbulent years. In 1984, a group of senior managers, led by president Dennis Vollmershausen, son-in-law of co-founder Wes Magill, along with a California investment group, Sequoia, acquired the company.

Meanwhile, in Finland, a group of forest machine companies was being assembled by Rauma-Repola, a diversified manufacturing and forest company which had become active in the forest equipment market in the early 1970s with the acquisition of the venerable Tampere manufacturer, Lokomo. Next, in 1980, Rauma-Repola purchased Ösa, Sweden's primary forest equipment maker, in a two-stage transaction which caused great consternation among proud Swedish machine makers. At the same time, negotiations were underway with Kockums. However, before a deal could be concluded, it was decided thinning machines were required to round out the product line of big harvesters built by Ösa and Lokomo. Attention was focused on Forrex, which had earlier acquired Makeri and Bruun, and had found great success with the Mini-Bruunett. In 1981, the Forrex stable joined Rauma-Repola, giving the Finnish firm a toehold in the US and the beginnings of a global presence, which was expanded later in the year through a joint production agreement in Brazil with the American company, Beloit.

By this time, a merger frenzy had developed on both sides of the Atlantic. Rauma-Repola and Valmet entered into a brief, unconsumated courtship, and then Kockums came back into the picture. In 1984, the Swedish company — along with the president of its Finnish operations, Mikko Rysä — was brought under the Rauma-Repola umbrella, where the growing group of forest equipment companies had been reorganized into Lokomo Forest Oy.

At about this time, Rauma-Repola opened discussions with Eaton, which still owned Timberjack. The talks ended and Timberjack went over to the Sequoia group. After struggling to survive the recession, Timberjack regained its financial feet sufficiently to buy Koehring Waterous in 1988, and the following year picked up the Oregon-based Peerless Corporation, a maker of chip vans, trailers and loaders. In the meantime, the Finns had come courting again, and in 1986 Rauma-Repola and Timberjack people began talking to each other directly in a series of discrete meetings. During the intervening period, Rauma-Repola's appetite for a distant alliance had been whetted by inconclusive discussions with the giant Caterpillar and a joint arrangement with the Japanese excavator manufacturer, Komatsu. A marketing agreement was the sole accomplishment of the second round of talks between Timberjack and Lokomo Forest. The latter, by 1988, had acquired control of France's biggest forest machine company, Cemet-Agrip, and had changed its name to Forest Machine Group, or FMG.

By 1989, a market for Scandinavian harvesting systems was beginning to appear in North America and Timberjack was faced with the prospect of developing its own lines of cut-to-length harvesters and forwarders, or entering into some form of association with an established Nordic firm. Negotiations with Rauma-Repola were re-opened and a deal reached in June in Finland. The participants celebrated their new alliance, under the name FMG Timberjack, in typical Finnish style — with a sauna. They had just created the largest forest machine manufacturing company in the world by merging two quite different cultures, as reflected in their distinct timber harvesting systems. In 1994, the FMG designation was dropped and the Timberjack name retained to identify the group. In 1996, Timberjack bought Hudson Sales, a 32-year-old Birmingham, Alabama, manufacturer of delimbing machines, and Plustech, a Finnish research and development company. The well-known donkey symbol, which identified Timberjack machines since the company's inception, was adopted by all divisions of the group.

This merger added a new element to North American logging practices. At the time of the purchase, about ninety percent of timber harvesting outside Scandinavia was done utilizing full-tree systems. In the Scandinavian countries there had been a major shift — about eighty percent of the total harvest — toward a new, cut-to-length logging system, utilizing rubber-tired feller processors or harvesting machines, which produced accurately bucked and delimbed logs at the stump, and rubber-tired forwarders to haul them to roadside.

“

*In the early 1980s, we were negotiating the sale of eleven Lokomo 993s to Calabari, Nigeria. The sale was brokered by the big British financing and consulting company Parson's Whittemore & Lyddon. Their representative was Ian MacLeod-Smith. Ian was a traditional British gentleman, so we called him Lord.*

*Chief Okon, a local tribal chief from Nigeria, was sent to examine our technology and suitability. We were there to greet him when he arrived at the small Härmälä airport in Finland on a DC3. The man was huge, a real king-size tribal chief, dressed in patent leather shoes and a thin poplin Ulster coat. It was February and freezing cold, so he was shivering when he arrived.*

*We spent the whole day showing him the factory and machines stored on the grounds. We were so eager that we didn't notice that Chief Okon had not said a word for hours. The Chief was bluish gray. And his patent leather shoes were full of snow.*

*That evening we took Chief Okon to a sauna in the Tammer Hotel. I stayed in the dressing room to chat and sip beer with Lord. Suddenly I remembered that the Chief had been alone in the steaming hot sauna for over half an hour. Alarmed, I rushed back to the sauna. The lighting was dim and the walls were blackened, so at first I couldn't see a thing. But then, back in the corner, I saw the shining teeth of a very wide grin: "Raul, I am feeling good; for the first time since I left Nigeria I feel like home!"*

Raul Wisotzky

During the latter stages of the Timberjack-Lokomo romance, another trans-Atlantic relationship evolved. In 1988, the Gafner family sold their Michigan-based operation to the Finnish state-owned Valmet Corporation. The Gafner enterprise had been started by Emil Gafner in 1946, and established itself as a major player in the forest equipment business with the introduction of its pulp wood forwarder, the Iron Mule, in 1962. Valmet's first purchase in the US was in 1990 when it bought the Clark Ranger skidder line, which it sold to Oregon-based Allied Systems in 1993. In 1994, when Valmet and Sisu Inc., also of Finland, joined forces, the Sisu Logging Group was established, with the Gafner operation becoming Sisu Logging USA.

> "
>
> *Electronics are going to become as popular and make as big a change in this industry in the future as hydraulics did. Single-purpose machines are going to be not very popular. You're going to go to more dual-purpose machines, dual-function machines. They'll be more sophisticated. Electronics will make a major change in industry. To solve the problems that we're facing in industry with the environment we've got to have soft footprints. You can't penetrate the ground or break it. With smaller sales and smaller job sites, you need more mobility. You're not going to see much clear cutting and so you're going to have to develop machines that don't skin up the trees, don't knock over the undergrowth, can work both in pre-commercial thinning and in select cutting a forest. What does that mean? That means that the equipment has got to be lighter, more mobile and versatile. That's where I see the industry going.*
>
> *You're going to see much more control of the power so that we don't waste a lot of fuel, where the engine is not controlled by the operator but by whatever function is needed, and only the power required. Not pushing oil through the system continuously, but only what is needed. That type of thing.*
>
> Len Vizina

The global mergers provided the North American partners in these arrangements with a line of single-grip harvesters and compatible forwarders developed by their Finnish and Swedish counterparts during the 1980s. This equipment was designed specifically for selective, low-impact harvesting of timber. On a global scale, the new equipment, in combination with tree-length systems developed in North America, allowed Timberjack and other globally-oriented firms to provide customers around the world with a complete range of harvesting equipment. For North American loggers, faced with increasingly successful demands from a growing preservationist lobby, the cut-to-length system offered a way to harvest timber in sensitive sites that was acceptable to regulatory agencies and the public. With relatively minor adjustments, Scandinavian systems could be adopted in North America, without the necessity of going through a long and expensive research and development process, and sales of cut-to-length machines soared.

The new Scandinavian machines had several characteristics, apart from their basic harvesting capabilities, which made them attractive on North American and other markets around the world. The automated, computer-based measuring and control equipment built into harvesters provided a capability beyond the ability to efficiently cut logs accurately. Initially, these devices were capable only of measuring log lengths. In the mid-1980s they gained the ability to measure log volumes, eliminating the time-consuming task of scaling. Eventually cellular and satellite telephones began to appear in forest machines. In harvesters, they were quickly used to plug the harvester's measuring system into a network which included log-supply people, head offices, mills and beyond to sales and marketing personnel. In theory, it became possible to reconfigure the processing program on the harvester in immediate response to orders received by distant lumber salesmen.

The complexity and precision capability of many of the machines developed over the past decade have enormously increased the level of skill required of operators. As recently as fifteen or twenty years ago, logging was looked upon as an unskilled occupation, paying low wages and often attracting the least capable members of the workforce. Now it is rapidly becoming a highly skilled job, paying well and attracting sophisticated workers. Special, rapidly changing skills are needed for an operator to fully utilize the capabilities of the new forest machines. Learning these skills has become a major task, far more complex than the old method of sending a new worker out into the woods for an hour or two of instruction from an experienced hand. On the job learning with a half-million-dollar machine is not an option.

One response to this challenge was the development of a simulator by Plustech. The equipment resembles a golf simulator, but the club is replaced by harvester controls and the video-projected golf fairway becomes a forest. The operator practices with a simulated felling head in a virtual forest. The equipment offers experience in controlling the felling head and processing trees into logs. It instructs, points out mistakes, and monitors the operator's development, indicating when he is ready to take over the controls of a real machine. Another application for the simulator is the planning of forest landscapes and harvesting. A database contains a digital map created with the help of aerial photos, topographical maps and forest inventories, and is enhanced with an image library of trees. Logging contractors and forest owners could, it is projected, develop a harvesting plan that would show the owner what his forest would look like after it was logged.

An inherent characteristic of the new forest machines is a high level of industrial design. Previously, machine design derived from the cost-efficient decisions of mechanical engineers. Industrial designers became part of the forest-machine engineering teams in the early 1990s, beginning in Finland where the reverence for good design is almost a national religion. Their assignment was to streamline the machines for emotional and practical reasons. On the emotional side, their task was the same as a car designer's: To make the machine stand out from its competitors by communicating a balance of power and gentleness. Even though a forest machine is a tool, the thinking went, its design gives the owner pride in his machine and dedication to his work in the forest.

On the practical side, the influence of industrial design is most noticeable in the cab. Today's cab is a spacious and ergonomically sound work environment. Noise and vibration levels have been reduced to a fraction of what they were. Heaters and air conditioners maintain a comfortable working temperature and clean air. Visibility has been improved with curved panoramic windows. The evolution tends toward the plastic bubble of a helicopter cockpit.

The timber harvesting systems that have emerged in the mid-1990s have dispersed rapidly. As recently as twenty-five years ago there were a great variety of logging systems and logging machines used throughout the world. Today, there is a globalization of equipment, a standardization of machines wherever mechanization has occurred. Almost everywhere logging operations have been fully mechanized, either a full-tree system utilizing feller bunchers and grapple skidders or a cut-to-length system is utilized. On the limited number of sites where these two systems cannot be used — very steep slopes, for instance — mobile cable yarders are generally employed.

As forest industries are modernized in more and more countries such as Chile or Indonesia, or where forestry is still emerging, as in Siberia, the need for mechanization rapidly becomes apparent. The highly sophisticated harvesting equipment which has evolved, almost entirely in Scandinavia and North America, is available in the developing forest economies without them having to go through a long development process. Because of the versatility of many of these machines, built to deal with the diversity of North American and Nordic conditions, they require minimal modification to meet the needs of loggers in most of the world's forests.

“

*How to best look after the forest. That's the driver right now, as I see it. I don't think we'll let the harvesting cost go up, but I think that's the challenge to the manufacturing sector. I look for the next ten years to be really exciting. I think we have had a very exciting period until now, but I think the next ten years is going to be even more exciting. In some way I wish I was a younger guy. But I really do believe that it's going to be a really exciting time because we have to solve the problems.*

LEN VIZINA

"

*In the late 1960s, through Timberjack, we were trying to get a harvester that could go into the swamps. The idea was being able to stay on top of the ground, float, and not destroy the ground and be able to move in low-ground areas. By this time the whole industry was using skidders. The horse had been phased out by then. We wanted a harvester to cut the tree down, delimb it right at the stump and put it into a bunch on the back of the harvester and then move along to the next tree, so that eventually we would have a cord, 15 or 20 trees, in a bunch on the back. We'd put a choker on it and drop it. Then a skidder would come and hook onto that choker and pick it up. This was going to be our mechanized way of doing things.*

*We were hoping to get our man-day productivity up from about five cords to about 10 or 12 cords. So instead of having a three-man crew, where you had a man felling the tree, a skidder skidding it and another man delimbing it, we had a two-man crew producing twice as much wood. The concept was great. They just didn't have the right materials, the alloys at that time.*

Garnet Bell

Another aspect of this equipment which is unique in the history of forest machinery is that, increasingly, it is no longer looked upon as logging equipment, but is seen as forest management machinery. Forward-thinking owners and operators of both cut-to-length and full-tree systems are adopting an outlook and a set of logging practices which give more consideration to the forest they leave behind than to the timber they are harvesting. An operator with this kind of an attitude does not consider a feller buncher a machine to log trees so much as a tool to increase the value of the forest stand in which he is working.

Where the innovative focus of forest-machine design will lie in the future is an open question. It is quite possible that new types of machines to work in difficult sites will appear, such as two now in the prototype stage. The Forestry and Forest Products Research Institute at Ibaraki, Japan, has built the Tri-Track Mover to operate as a prime carrier on that country's steep, fragile slopes. It consists of three pairs of tracked units, each equipped with a diesel engine and hydrostatic drive mechanisms. The three units are connected with height-adjustable arms which support the machine's working gear and which swivel, enabling the machine to remain level on 60 percent slopes, while crossing over obstacles, such as rocks, stumps or tree seedlings, up to 1.3 meters tall. If the Tri-Track Mover signifies the beginning of a long-overdue Japanese participation in the global forest-machine manufacturing business, the future of the industry will undoubtedly hold many surprises.

Perhaps the most dramatic new machine now in existence is the Walking Machine built by Plustech in Finland. The idea of a walking machine is not new, and an earlier version, the three-legged Menzie Muck, first appeared in Switzerland in the late 1970s. Intended initially as a carrier for a backhoe, it was equipped with an experimental felling head to mechanize timber falling on steep slopes. Because its motion was controlled by the operator, it was slow and cumbersome, and disappeared as a contender.

The Plustech Walking Machine, is a radically improved version. Its six legs are computer controlled. The operator simply decides on the direction of travel and the computer controls the motion of the legs, which are equipped with sensors to detect unsuitable ground. The purpose of the machine is to serve as a carrier for a harvester head or other forestry tool on terrain too steep or sensitive for wheeled or tracked vehicles. The machine's "feet" can be fitted with a variety of surfaces to traverse different ground conditions — a large surface for swampy areas, for instance, or long caulks for climbing extremely steep slopes. Still a slow-moving vehicle, the Walking Machine is a few years away from commercial production, but already is exciting machine designers and builders around the world.

The Walking Machine, probably the most sophisticated piece of logging equipment ever built, appeared just over a century after the first mechanically-powered logging machine, John Dolbeer's steam donkey. During that time much has changed in the world, including the forests themselves, the machines used to harvest them and human attitudes about them. It is curious, and somehow fitting, that the image of a tenacious, hard-working creature, the donkey, is associated with both machines — in Dolbeer's case with the name of his machine, and for the Walking Machine with the corporate logo of Timberjack.

The Mini Bruunett with a brushing head clearing under-vegetation at a birch plantation in Sweden in the 1980s.

The Silva Nova planting machine was developed in Sweden in 1983 following eight years of intensive work. Although it was one of the more successful planting machines, the problem of mechanized planting still awaits a solution.

The SP 21, a falling head designed jointly by machine builders and logging contractors, accelerated the development of single grip harvesters in the early 1980s. It felled, delimbed and bucked.

Valmet quickly responded to the challenge presented by the SP 21, and came out with the 935 harvester head in 1984.

Valmet's 901 single grip harvester with a 940 felling head.

The massive Koehring-Waterous feller forwarder demonstrating its capabilities at Thunder Bay, Ontario, in the 1980s.

The three-wheeled Mor-Bell faller evolved in the US from a South African sugar cane loader. The builders of this machine, Morbark and Bell, each went on to manufacture their own three-wheel feller bunchers.

The Lokomo 990/910 system was one of the most popular logging systems in the late 1980s.

The Rottne two-grip harvester, the Snoken, introduced rubber feed wheels to eliminated damage to logs caused by spikes. Because of the location of its delimbing and bucking head it was called a 'tail processor.'

Valmet's 882 forwarder from the late 1980s. A no-frills machine, it had a turbo-charged engine which provided increased pulling power in relation to total weight.

The Valmet 911 single-grip harvester from the 1990s is used for both regeneration harvesting and thinning.

The Timberjack 600 series feller bunchers were introduced in late 1980s. The drowbar pull close to machine weight gives the tractive effort to maneuver easily over rough, soft or hilly terrain.

The evolution of skidders has been enourmous in the past four decades. The 90s machines include forward-weighted balance and improved power-to-weight ratio. The new cabs are ergonomically designed. Pictured here the Timberjack 460 skidder which was launched in 1996.

The small single-grip harvester of the early 1990s, the Timberjack 570, a descendant of the Lillebror. Rubber wheels covered with chain mesh have replaced steel spikes on the harvester head's feed rollers and do not damage timber.

Ponsse's Ergo Series single-grip harvester in the 1990s. Ponsse was the first forest machine manufacturer to receive the ISO 9001 quality certificate.

The Hydro-Ax feller buncher is a US-built, hydrostatically driven, drive-to-tree machine developed by Len Vizina. It is widely used in plantation forests in the southern US.

The Morgan SilvaCom skidder evolved in the 1990s from a carrier developed for the Reforester tree planter and is the first hydrostatically driven skidder. It is built in British Columbia.

Designed and built by former logger, Pat Crawford of Wisconsin, the Timbco feller buncher is designed for selective harvesting on steep slopes. The cab and boom platform levels in all directions.

The Timberjack 1270B harvester has a computerized measuring and control system which sets timber harvesting parameters based on the owner's priorities. The system monitors production by volume and species.

The ergonomically designed cab of a modern forest machine is equipped with air-conditioning, a mobile phone, a stereo system and computer display terminals. Everything is controlled with joy sticks and touch buttons within reach of the operator's finger tips. Darkened polycarbonate windows reduce heat radiation and eliminate glare.

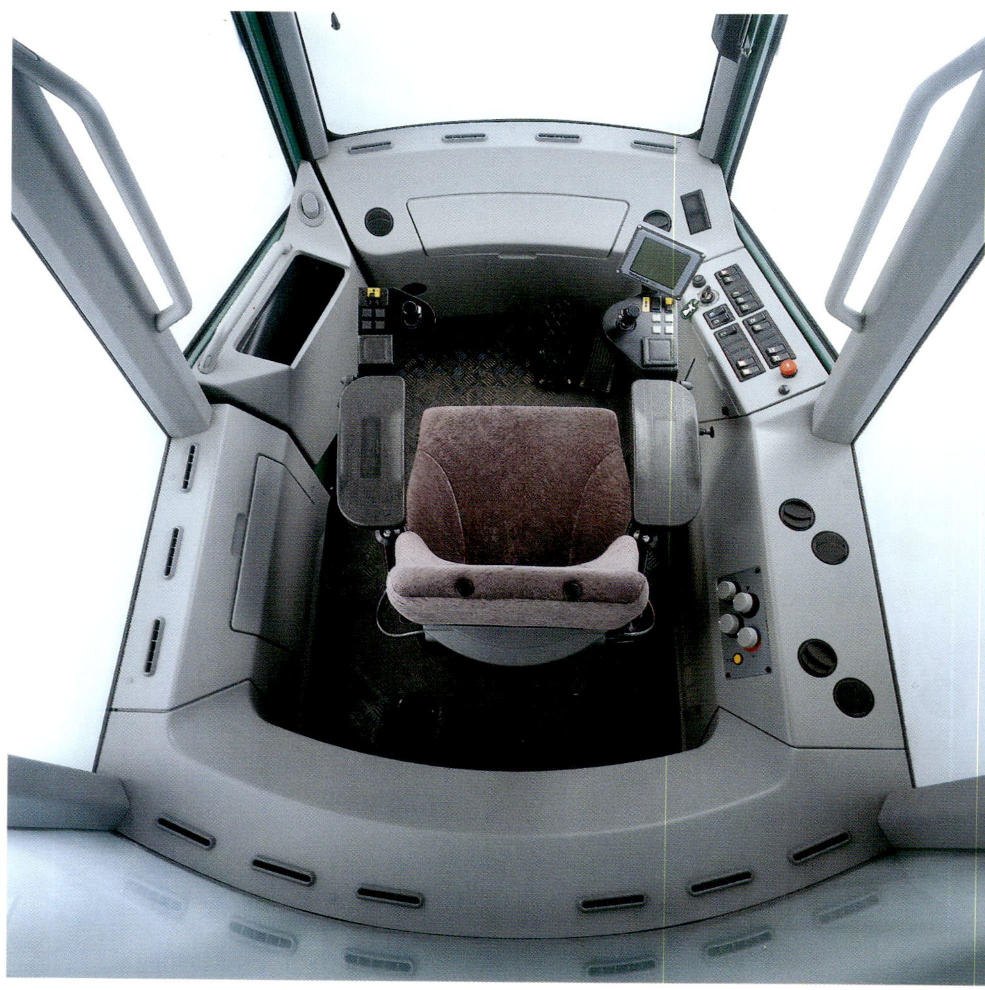

Today's forwarders have control systems for the power transmission and loader; they are adjustable according to the operators preferences and inclinations. The computer-controlled power transmission is under development. It will control each of the drive wheels and adjust the wheels' ground contact. Pictured here the Timberjack 810B forwarder.

Plustech's six-legged Walking Machine, equipped with a single-grip harvesting head, moves gently through a forest on feet resembling those of an elephant. This prototype is being developed to navigate terrain inaccessible to wheeled or track machines.

The prototype Tri-Track Mover has been developed by Japan's Forestry and Forest Products Research Institute as a prime carrier to operate on steep, rough slopes. The independently-powered track units are attached by hinged arms and enable the machine to straddle rocks, stumps and other obstacles.

# Epilogue

The evolutionary changes in forest machines wrought over the past century provide only a few clues about how these machines might develop over the next hundred years. Something we have learned from looking at the history of logging is that entire lines of equipment may disappear, almost overnight, for reasons that have nothing to do with their inherent capabilities. For instance, the various machines which made up the steam-powered railway logging technology which evolved in North America during the late nineteenth and early twentieth centuries were displaced rapidly by new kinds of machines powered with internal combustion engines. Likewise, dozens of mechanical devices utilizing cables — such as loaders and dozers — were supplanted in the 1950s by hydraulic systems. We cannot predict these changes, where even small technical innovations can give an immense advantage to one type of machine over another. Someday, a new form of mechanical muscle, perhaps utilizing a new generation of pneumatic equipment, might replace hydraulics and induce a substantial evolutionary diversion. Perhaps such an invention would convert the Walking Machine into a fleet-footed runner capable of bounding lightly through the forest bearing a load of logs.

We have also seen how events external to the forest industry can dramatically influence its development. Major historical events such as the Great Depression and World War Two suspended development of logging machines for twenty years. After the war, a period of rapid evolution occurred in which numerous new machines appeared incorporating technical developments — hydraulics, alloys, electronics — achieved during the war. The quick success of the McCulloch chainsaw, which was built using wartime innovations, is an example. Similarly, severe geopolitical or geophysical events — for instance, a breakdown in civil society or a large meteor striking the earth — would have dramatic consequences for the world's forest industries and the machines they require. We cannot foresee these catastrophic events, only anticipate that, in some form or another, they may occur.

One of the characteristics of people who work in the logging industry is their ability to adapt to unexpected change. In a business such as logging, where things often fail to occur as planned, this trait is necessary for success or survival. There are no two trees on the planet which can be logged in precisely the same manner, a condition calling for a innovation and flexibility in the application of logging technology. It fosters a way of looking at the world which is illustrated by a warning to travellers erected on northern Vancouver Island in British Columbia in the 1960s. At an important junction of major log-hauling roads, loggers crushed an automobile beneath a huge Western red cedar log, three metres in diameter. Then they posted a sign: "EXPECT THE UNEXPECTED." It is good advice to keep in mind when contemplating the future of logging and the nature of the machines which will be built to work in the forests.

“

*The world's population will continue to grow in the medium-term future. In 2010 it is projected to reach 7.1 billion people compared to 5.2 billion in 1990. About 6 billion will be living in developing countries where there will be 1.6 billion people more than at present. In the developed countries changes will be marginal. The global demand for resources will dramatically increase in the tropics, not only because there will be more people but because they will be wealthier as well. Demand growth in forest products will take place mostly in Asia accounting for more than a half of the world's population in 2010. To meet the future demand for forest products represents a major opportunity to forest industries.*

*The extreme effect on forests of population pressures is their devastation. At present, conversion of forest land in developing countries is estimated at about 12.9 million hectares, most of which is needed to meet the growing need for land for agriculture and grazing, while in industrialized countries the forest coverage is increasing at an annual rate of .12 percent. In 2010, there will be .2 hectares or about 30 percent less forest area per capita in developing countries than at present. This is even more alarming when it is known that the present level is already low and in many areas grossly insufficient to meet the demand for forest products and services by local populations. There are also sighs that logging — both directly and indirectly — increasingly contributes to forest degeneration in many areas, casting a shadow on the acceptability of harvesting of industrial roundwood.*

Markku Simula

Perhaps the most important factor for the future of the industry is population size. If current trends continue, the world's population will increase dramatically — by almost 100 million people a year — mostly in developing countries. These additional people will need an additional 60 million cubic metres of wood each year — for fuel if they remain poor or for conventional forest products if they improve their economic conditions. More and different kinds of forest machines will be needed to harvest the increased timber harvest. Under this scenario the global forest economy will retain a fundamental characteristic of the past century — constant economic growth and expansion. On the other hand, it is impossible to conceive how such increases in either population or timber harvesting could be sustained for another century. If the world's population does not increase as projected — because of catastrophic events such as war, disease or famine, or as a consequence of a consensus to curb growth — then unrestrained economic expansion will not provide the incentives it has in the past. The evolution of forest machines in a steady-state forest economy will be much different than in a climate of constant growth.

Changes in the environmental movement will also have a profound effect on the character of the world's forest industries and machine makers who supply the equipment. It is doubtful that the intense scrutiny under which the forest industry now operates in some parts of the world will be sustained. Not that the public and their political representatives will tire of their mission to impose a reconfiguration of the ways in which forests are used and leave foresters and loggers to their own devices. Rather, they will establish a new set of parameters within which the industry will have to function. The precise nature of these parameters is not yet clear, but whatever they are they will constitute a new set of assumptions about forests leading to a need for new types of forest machines.

To appreciate how this might occur, we need only look at the consequences of the North American conservation movement which arose about a century ago. It was as passionate and as powerful a movement in its time as the environmental movement is today. It was opposed to the prevalent assumption of the time that forests were an impediment to civilization and should be permanently removed. The agent of this removal was the nineteenth-century logging industry — with Paul Bunyan as its archetypical hero — and the nature of this attitude heavily influenced the design of logging equipment. The trend was toward large, powerful machines which cleared the land of trees in huge, unbroken clearcuts. The impact of these machines on forest ecology was not a concern since the object was to liquidate the forests. The original forest in the US South was a victim of this attitude and the technology it spawned. The conservation movement which appeared in the 1890s, championed by professional foresters and a few forward-thinking timber operators, opposed this idea with another — the efficient use of resources obtained from a permanent forest. Paradoxically — by current standards — the emergence of a forest industry with an interest in perpetuating itself assured the survival of North American forests. By World War Two, the principal of sustained-yield forestry was widely accepted in North America, leaving a new set of parameters concerning forests within which a reorganized forest industry thrived. This shift in thinking, from an acceptance or pursuit of forest liquidation to a regime of sustainable forest management, is taking place now in much of the developing world.

The legacies of the contemporary environmental movement are still in the making. The uncertainty of the outcome is probably due, in part, to the confusion and differing attitudes about basic forest issues. For instance: What is a forest? Quite clearly a treed area in the Amazon which has never been logged is a forest. But what about a treed area in the American South, which was allowed to revert to tree cover after it had been cleared in the early years of this century and used for decades to grow cotton or tobacco? And if that is a forest, what about an exotic eucalyptus plantation on land in Portugal that never, in human history, has been covered with trees?

Are plantations even forests? What is becoming clear is that plantations can supply us with substantial portions of our industrial timber requirements much more efficiently than natural forests. But is the fact that they are covered with trees sufficient to classify them as forests? And if it is, does that mean they will need to be treated within the same set of parameters that will govern the use of natural forests? Will the people who manage them be able to operate within the different parameters which apply to agriculture? Or, on plantations, will there be a merging of the historically distinct enterprises of agriculture and forestry? Another possibility is that plantations will be so successful in supplying timber that they will inhibit the expansion of timber harvesting into the natural boreal forests of northern Canada and Siberia and the untouched tropical rainforests of the equatorial region. There is no consensus among foresters, loggers or environmentalists about these questions so it is difficult to guess now how they will be resolved.

“

*I do not see major machinery changes in the next ten years. Beyond ten years, major improvements should be expected. Laser technology may be employed to fell and process trees. Hovercraft may be the low-impact machines of the future which will move trees from stump to roadside. Timberlands will be much more regulated by foresters, that is, scheduling of harvests and regeneration will be much more precise and accurate which will allow machinery to be designed for more specific tasks rather than for all tasks.*

Bob Lanford

> *While it appears that forest machine development and industrialized mass production is primarily centralized in North America and Scandinavia, there are new companies cropping up all over the world, wherever timber is harvested. Innovations are emerging in small workshops, while sitting on a stump and thinking, and in technology centres. A good example of the latter is the Tri-Track Mover developed in the Forestry and Forest Products Research Institute in Ibarake, Japan. Designed to operate on the steep mountain slopes of Japan, the machine has three independent, interlinked units, each with its own pair of tracks, diesel engine and hydraulic transmission. Thanks to the design of its chassis, the Tri-Track Mover is capable of agilely slaloming between trees and crossing over boulders, stumps and trees up to 1.3 metres high. The machine is still in the prototype stage, but it is proof that surprises can come from any direction.*

Whatever the outcome of these issues, and others we cannot yet imagine, they will heavily influence the design and function of the forest machines which appear in the twenty-first century. This evolutionary process takes us into the very essence of the industry. The building of forest machines and their use in forests requires and fosters a certain kind of inventive genius. One of the most obvious and curious historical facts about this business is the large number of machines which have been conceived of, designed and built by loggers — most of them lacking formal education in the complex array of skills required for such a task. Now, there is a new generation of logging machines requiring highly skilled and capable operators who, in the coming decades will be even better equipped for the task of finding mechanical solutions to the problems raised by shifting parameters and diverse physical challenges.

A relatively constant factor in the future will be the physical conditions in which forest workers perform their tasks; the forest landscape will remain more or less the same. This indicates some directions in which developments will proceed. On easy ground conditions we can expect to see the extension and refinement of the existing cut-to-length and full-tree harvesting systems. Improvements in these systems will continue to reduce the impact of logging on the forest, to increase the value of the timber harvested and to increase worker productivity. Because developments in this sector are already well ahead of need as it is perceived in most of the world, there is not an enormous requirement for new kinds of machines to harvest this type of terrain, even though opportunities may increase for use of the systems already developed.

Where there is a greater need, and where some new or vastly improved equipment may appear, is on the more difficult logging sites — steep slopes, swampy areas and inaccessible stands. Some of these equipment requirements are already defined — for example, agile carriers, such as the Walking Machine, are needed to mechanize falling and processing in steep or wet areas inaccessible to wheeled or tracked carriers. To move logs cut in these areas to roadside, continued development of cable logging systems incorporating electronic control systems and other components to reduce labour input is likely. So is the increased use of aerial equipment such as balloons and helicopters. At some point there may be sufficient timber available in otherwise inaccessible locations to justify the development of a purpose-built logging helicopter, perhaps an advanced version of the Russian Kamov KA-32. Or the discovery of improved, impermeable fabrics or a new propulsion system may increase the capability of free-flying balloons and enable their adaptation to logging. The development of hovercraft forwarders for use in wet areas is also a possibility. Because valuable timber will continue to grow on difficult sites, we can be reasonably sure someone will build a machine to harvest it within whatever constraints are imposed.

> “
> *The shortwood method has proved to be highly suitable for the harvesting of plantation forests. Another reason for the spreading use of the shortwood method is the strong technological development pattern which is presently concentrated on harvesters, forwarders and loaders. The tree-length method is expected to remain the major method in the harvesting of natural forests with tall and large-diameter trees. However, the harvesting of this type of forests will have a decreasing role, as the old-growth forests in the boreal and temperate zones are currently a scanty resource and the utilization of the remaining tropical rainforests will be strongly restricted for environmental reasons.*
>
> MARKKU SIMULA

Perhaps some of the most dramatic advances in equipment design will be machines built for tending and harvesting plantations. Today, there are about 100 million hectares of tree plantations in the world, with more on the way. Not only will new kinds of machines evolve to tend and harvest these lands, but the machines used will be automated to a higher degree than other logging equipment. It is in the plantations that robotic logging machines will probably first appear. The mating of robotic technology with existing forest technology should produce a wide range of new and interesting applications. Plantations will also foster a merging of forestry and agriculture, in terms of the crops grown, in regard to the people and firms who build equipment, and in the nature of the machines themselves. The melding of logging and farming, common prior to World War Two, will create a synergetic climate conducive to innovative machine design. The evolution of the Bell tri-wheeled sugar cane loader from South Africa into a feller buncher may be a precursor of this type of development.

What may also emerge are multi-functional machines which combine forest-tending tasks with timber harvesting — harvesters which also prune during a thinning operation, or forwarders equipped with mechanized planting equipment. The possibility of managing forests for additional, non-timber values — for example, food, wildlife, water, recreation — as well as for timber production may provide opportunities for new types of economic activity, including the building of new forest machines. The tending of forests as a productive enterprise is an activity gaining momentum, and for the most part this work is not mechanized. It is an obvious field of opportunity which is unlikely to be ignored for long as, increasingly, forest management is perceived as a productive enterprise.

At the end of the twentieth century, the future of the forest industry often appears uncertain, at least in comparison to previous periods. Throughout much of the industrialized world there is an intense debate about future use of forests. At one end of the spectrum are extreme environmentalists who would shut down the global forest industry and prohibit the use of forests. At the other end are besieged defenders of the industrial status quo who, like the post-war Nordic horse loggers, perceive all change as a threat to their existence. It is highly unlikely that either of these entrenched positions will prevail over the course of the twenty-first century. Instead, it is probable that a fusion of these views will evolve based on the recognition by all that civilization cannot survive without the continued existence of the world's forests, and their use by humans. In such a world, the forest industries, the people who work in them and the machines they use to perform their tasks are essential.

Exactly how events will unfold is anybody's guess. The Czech author and politician Vaclav Havel has observed that we are heading into a future where anything is possible and nothing is certain. That judgement applies as much to the evolution of forest machines as it does to the fate of mankind.

# Appendix

# Glossary

**aerial logging** method of moving logs using air-borne equipment such as a helicopter or balloon

**alligator** an amphibious warping tug used to tow logs on water, which can also winch itself over land

**arch** a device towed behind or mounted on a vehicle to elevate one end of logs during skidding

**articulated** refers to a vehicle, such as a skidder or forwarder, hinged at the centre for steering

**ballhooting** a method of hand logging in which logs are slid or rolled downhill

**beavertail bar** a chainsaw bar, wider on the outer end than the motor end, and lacking a helper's handle

**big wheel** various types of two-wheeled log haulers used in dry forest areas, ranging from two to four metres in diameter; the logs were raised under the axle; also called "whims" in Australia

**board foot** a lumber measurement of 144 cubic inches; usually expressed as a board one foot square and one inch thick

**board foot log rule** indicates the number of board feet which can be sawn from a log using standard sawing practices; the log volume needed to produce a volume of lumber; commonly used to measure sawlogs in the past, now has largely been replaced by cubic measurement

**bobtail truck** a single-axle truck used for hauling logs, particularly pulpwood, without a trailer; primarily in the southern US

**bogie** a wheeled or tracked vehicle undercarriage with a set of wheels or sprockets on each side that are mounted on either end of a beam which pivots on the axle, to more easily traverse uneven ground

**buck** to cross-cut a tree into shorter lengths, or logs

**bullbucker** the person who supervises fallers and buckers

**bummer** a low cart, tracked or wheeled, used for carrying logs on top of the axle; also called a "jinker" in Australia

**butt** the bottom of a tree, or stump end of a log

**cable logging** the skidding of logs using wire ropes and winches

**capstan** a spool or drum used to pull a rope or cable by friction without being attached to it

**chance** a logging operation; also called a "show"; also see "side"

**choker** a short length of chain or wire rope with attachments used to attach logs to a skidding line or machine

**clam bunk** a set of hydraulic clamps, which open at the top, mounted on the rear of a vehicle to hold one end of logs for skidding

**clearcut** cutting all the trees in an area; in some countries implies permanent forest renewal; see "regeneration harvest"

**cold deck** to pile logs for future transport

**cord** a unit of wood volume measurement representing a pile of logs measuring four by four by eight feet, or 128 cubic feet of wood, bark and empty space between pieces

**crawler** a tractor running on continuous tracks instead of wheels, commonly called a Caterpillar after most well-known brand

**crosshaul** loading logs with a cable by passing a line around a log and pulling it onto the load; also see "parbuckle"

**cut-to-length** logging systems in which felled trees are processed into log lengths at the stump before they are carried to the road or landing; an alternative system to "full-tree" logging

**debark** to remove bark from logs

**delimb** to remove limbs from logs or trees

**directional fall** to fall trees in a predetermined direction

**donkey** a wide variety of machines, all utilizing one or more cable winches and lines to skid logs

**duplex loading** a donkey with two independently operated winches and lines used to load logs

**faller** a person who cuts or falls trees

**feller** a device for cutting down trees, sometimes used instead of "faller"

**feller buncher** a machine to fall trees and place them in piles or bunches on the ground

**feller forwarder** a machine to fall trees and transport them by carrying them completely off the ground

**feller skidder** a machine to fall trees and transport them by dragging them with one or both ends on the ground

**flume** a water-filled trough down which logs or lumber are floated

**fore-and-aft road** a truck road built of logs hewn flat on top and laid end to end on a log trestle

**forwarder** a machine to transport logs or trees by carrying them completely off the ground

**frame steering** a method of steering a vehicle; the frame is built in two parts, each with a set of fixed wheels, and attached with hinged pins, enabling it to bend with the aid of hydraulic rams; also known as "bent-frame steering" or "articulated steering"

**full-tree** the entire upper portion of a tree, consisting of trunk, branches and leaves or needles; same as "whole-tree"

**full-tree logging** logging method in which felled trees are transported to roadside or a landing for subsequent processing and transport

**gate** a steel grid through which full-tree logs are pushed to remove their limbs, primarily in southern US

**grapple** a hinged set of jaws, activated hydraulically or with cables, to grip logs for skidding, yarding or loading

**grapple yarder** cable yarding using a grapple instead of chokers to secure logs

**grapple skidder** a vehicle which uses a grapple to assemble and hold logs for skidding

**gyppo** a small logging operation which, in the past, survived by cutting corners or "gypping;" in later times refers to a small operation which gets by on ingenuity and hard work

**half-track** a vehicle with a set of driving tracks on the rear and steering wheels on the front; various versions, such as Alfta tracks made in Sweden, were commonly added to farm tractors or trucks to give them added mobility in the forest

**handlogging** cutting trees and moving them without the use of powered machines; known as "ballhooting" in the US, "kuda-kuda" in Sarawack, "centipede" logging in Australia and "chienning" in Central Europe and Quebec

**harvester** a machine which falls trees and performs processing functions at the stump

**haulback line** in cable logging systems a cable used to pull the main or hauling line to the setting so logs can be attached

**heel boom** a loading system in which logs are grasped off-centre by tongs or a grapple and the short end is raised against the lower end or heel of the boom to provide controlled lift of the entire log

**high-lead** a system of cable logging in which the mainline passes through an elevated block, providing lift to the leading end of the log being skidded

**iron monger** loggers' term for an equipment salesman

**jammer** a cable loader with a spar or A-frame mounted on a sled or vehicle and occasionally used for skidding logs short distances

**jinker** see "bummer"

**joystick** a hydraulic control lever, used primarily for controlling direction of movement

**kedge** to move a vessel or log boom by setting an anchor ahead of the load and reeling in the anchor line; also known as "warping"

**knuckleboom** an articulated hydraulic boom that imitates the movement of a human arm

**mainline** the line on a cable yarding system used to haul logs into the yarder

**off-highway** usually in reference to trucks designed to handle loads in excess of highway limits, on roads which are not public highways,

**pallets** pipe or metal frames built to hold a specified volume of shortwood pulp logs, which can be loaded onto a truck or forwarder for transport

**pan** a sheet of metal with an upward curving forward edge on which the leading end of logs are laid to facilitate skidding

**parbuckle** a method of rolling a log or other object by passing a cable under it and back over top causing it to roll when the cable is pulled; also see "crosshaul"

**Peavey** a hand tool with a wooden handle, a steel point and a hinged, curved hook used to roll logs; invented by J.H. Peavey of Bangor, Maine

**processor** a machine which performs two or more functions on a felled tree, including delimbing, debarking, bucking, measuring or chipping

**pull boat** a barge or flat-bottomed boat equipped with a donkey to skid logs in swamps or other wet areas

**purpose-built** a machine designed and built to serve a particular function

**rear drags** a Scandinavian term for a small, two-wheeled cart; also see "bummer" or "jinker"

**regeneration harvest** Scandinavian term for a clearcut harvest prior to reforesting the area

**roader** a donkey used to skid a line of attached logs along a prepared skid road

**scarify** loosening of soil surface to facilitate reforestation

**second-growth** trees which grow after a first-growth or naturally-grown forest has been logged

**selective cut** one of several methods of partially harvesting a forest stand

**shear** a hydraulically powered scissor-like device for severing a tree in falling or bucking

**side** a logging operation, including the men and equipment needed to perform the required tasks; also see "chance"

**single-grip harvester** a machine with a single unit which fells, limbs and bucks trees; primarily used in "cut-to-length" systems; also see "two-grip harvester"

**skidding** moving logs by dragging them partially or totally on the ground with cable systems or vehicles; called "snigging" in Australia

**skidder** a machine for dragging logs partially or entirely on the ground; includes cable systems and tracked or wheeled vehicles

**skyline** a taut cable strung between a head spar and a tail spar, along which a carriage runs; a mainline lifts logs up to the carriage and pulls them and the carriage to the head spar

**slasher** a machine for bucking trees into logs

**sloop** a sled used to haul pulpwood or saw logs out of the woods

**snaker** a donkey used to drag logs along the ground

**snigging** see "skidding"

**spar** a wooden pole or steel tower used in cable logging

**spot** to position a rail car or truck in preparation for log loading

**sulky** a two-wheeled arch towed behind a crawler tractor which lifts the leading end of logs for skidding

**swing** to move logs from one skidder landing to another

**tail hold** the rear anchor of a line on a cable logging system, especially the anchor at the far end of a skyline

**thinning** the selective cutting of trees from an immature forest to stimulate growth of the remaining trees; called "commercial thinning" if the trees removed are of merchantable size and are removed for use

**tongs** scissor-hooks used for grasping logs for loading or skidding

**total-tree** the entire tree, including top, branches and roots

**tree-length** a log consisting of a full tree with the top and limbs removed

**two-grip harvester** a machine with one unit for falling a tree, and a separate unit for processing it

**warp** see "kedge"

**whole-tree** same as "full-tree"

**yarder** a donkey utilizing cables to move logs from the stump to a landing

# Wood Measurement Equivalents

| **Cord** | | **Cunit** | | **Cubic meter** (stacked) | | **Cubic meter** (solid) | | **Board Foot** |
|---|---|---|---|---|---|---|---|---|
| 128 cubic feet (4' × 4' × 8') | | 100 cubic feet of | | 1m × 1m × 1m of | | 1m × 1m × 1m of | | 1" × 12" × 12" FBM (Foot Board Measure) |
| apparent volume | | solid wood volume | | apparent volume | | solid wood volume | | approximate volume |
| **1** | = | 0.85 | = | 3.625 | = | 2.407 | = | 500 |
| 1.176 | = | **1** | = | 4.264 | = | 2.832 | = | 600 |
| 0.276 | = | 0.234 | = | **1** | = | 0.664 | = | 138 |
| 0.415 | = | 0.353 | = | 1.506 | = | **1** | = | 208 |
| 2 | = | 1.67 | = | 7.25 | = | 4.81 | = | **1,000 (MFBM)** |

# Oral History and Story Sources

**Bell, Garnet** a retired forester with the Spruce Falls Power and Paper Company; worked with Timberjack during the 1960s to develop a tree-length harvesting machine. Interviewed by Ken Drushka, 1996.

**Burch, Tom** a former Texas Logger of the Year and a woodlot owner near Jasper, Texas. Interviewed by Ken Drushka, 1996.

**Challenger, Dave** retired sales manager for Pacific Equipment Co., a major British Columbia chain saw distributor; grandson of pioneer truck logger George Challenger and son of pioneer chain saw developer Jack Challenger; owns one of the world's largest collections of power saws and other falling tools. Interviewed by Ken Drushka, 1991.

**Cliffe, Joe** started logging for Comox Logging in British Columbia in 1920 and retired from its successor, Crown Zellerbach, in 1961. Interviewed by Lynn Henderson, 1980. Source: Courtenay District Museum.

**Crawford, Mal** marketing specialist for Prentice loaders at Blount Inc. in Zebulon, North Carolina. Interviewed by Ken Drushka, 1996.

**Crawford, Pat** president of Timbco Hydraulics Inc. in Shawano, Wisconsin. Interviewed by Ken Drushka, 1996.

**DeBeck E.K., (Ned)** born in 1883, was a member of a pioneering New Brunswick logging family which began logging in British Columbia in 1867 and founded Brunette Sawmill near Vancouver in 1877. He later served as Clerk of the BB Legislature. Source: Unpublished memoirs, BC Archives and Records Service.

**Drake, Roger** founder of Franklin Equipment Co. in Franklin, Virginia. Interviewed by Ken Drushka, 1996.

**Eitel, George** began logging in 1896 at the age of 16 in Wisconsin. Interviewed by Elwood Maunder in 1957. Source: U.S. Forest History Society.

**Elofson, Hans-Åke** started at Volvo BM's marketing department in 1969, then worked for Ösa. Currently Marketing Analyst / Economist for Timberjack Group in Helsinki, Finland. Interviewed by Hannu Konttinen, 1996.

**Estes, Harold** President of Texas Timberjack in Lufkin, Texas. Interviewed by Ken Drushka, 1996.

**Fedje, Olaf** began falling timber in British Columbia in the 1930s and was a founder of Fedje & Gunderson, one of B.C.'s largest and oldest falling contractors. Interviewed by Ken Drushka, 1991.

**Fraser, Frank** began logging for the Riverside Mill and Lumber Co. near Eureka, California, in 1898. Interviewed by John Larson, 1953. Source: U.S. Forest History Society.

**Gustafsson, Gösta** veteran in the Swedish forest machine sales, marketing and exports, has worked for both Volvo BM and Ösa. Currently works for Timberjack's marketing support in Sweden. Interviewed by Hannu Konttinen, 1996.

**Hahn, Ray** began logging in Minnesota in the late 1930s, is the retired president of Hahn Machinery Inc. in Two Harbors, Minnesota. Interviewed by Ken Drushka, 1996.

**Helin, Esko T** author of the "First mechanized logging site" — an article published in the Finnish *Koneyrittäjä* magazine.

**Huusko, Pertti** joined the forest machine business in 1980 and has since then worked for Pöyry, Kockums, Ponsse, Lokomo Forest and FMG. Currently Timberjack's Vice President and head of the Harvesting Division. Interviewed by Hannu Konttinen, 1996.

**Kare, Erkki** has made a long career in forest machine development. At the moment President of Plustech Oy. Interviewed by Hannu Konttinen, 1996.

**Kelly, Tom** Woodlands Manager for Scott Paper in Mobile, Alabama, from 1978 to 1993. Interviewed by Ken Drushka, 1996.

**Kerr, Archibald** worked at many jobs in the British Columbia woods over a 60-year period retiring as a professional forester. Source: University of British Columbia Special Collections Library.

**King, John Leslie** born near Blackshear, Alabama, in 1895 and logged all his live in Alabama, Florida and Georgia. Interviewed by D.K. Knight, 1976. Source: Timber Harvesting magazine.

**Lanford, Bob** Associate Professor of Forest Harvesting at Auburn University in Alabama.

**Malmberg, Carl Einar (Cema)** started at the SDA machine-building station in the 1940s. Among other positions, he was the Technical Director of Skogsarbeten and is now retired in Solna, Sweden. Interviewed by Gösta Gustavsson, 1996.

**Moisander, Jukka** started out as a Product Development Engineer and is currently Product Support Manager for Timberjack's Harvesting Division in Tampere, Finland. Interviewed by Hannu Konttinen, 1996.

**Morey, Norval** began logging in Idaho in the 1930s, and southern Michigan after World War Two. He is the founder and Chief Executive Officer of Morbark Industries of Winn, Michigan. Interviewed by Ken Drushka, 1996.

**Norie, Frank** founder of Norie Brothers Logging and Vancouver Island Helicopters, both in British Columbia. Interviewed by Ken Drushka, 1991.

**Papunen, Kauko** works in marketing support for Timberjack AB in Tampere, Finland. Interviewed by Hannu Konttinen, 1996.

**Partington, Bob** a retired Director of Corporate Product Development at Goodyear Tire in Toronto. Interviewed by Ken Drushka, 1996.

**Peterson, Howard** began logging in 1933 for Crown Zellerbach, and worked for the company in Washington state for the rest of his career. Interviewed by Amelia Fry, 1966. Source: U.S. Forest History Society.

**Pinomäki, Sakari** an inventor, machine builder and Managing Director of S. Pinomäki Ky in Ylöjärvi, Finland. Interviewed by Hannu Konttinen, 1996.

**Pond, Leon** a son and grandson of loggers, grew up in a Canadian International Paper Company logging camp in Quebec during the Depression and Second World War. He retired from Weldwood Canada and works as a forest consultant in Vancouver, British Columbia. Interviewed by Ken Drushka, 1996.

**Simula, Markku** Ph.D. of forestry and international specialist on the economics of forestry and forest industries; President of Indufor Ky, a Helsinki, Finland based forestry consultancy agency.

**Small, Don** Timberjack's first staff employee, who later was shop superintendent. Interviewed by Ken Drushka, 1996.

**Stephenson, Ev** Group Leader in the Forest Engineering department of Union Camp Corp. at Savannah, Georgia. Interviewed by Ken Drushka, 1996.

**Stuart, Bill** professor of Industrial Forestry Operations at Virginia Tech in Blacksburg, Virginia. Interviewed by Ken Drushka, 1996.

**Vidgrén, Einari** started as a lumberjack and a driver, and is currently Chairman of the Board of Ponsse Oy, a forest machine manufacturer in Vieremä, Finland. Interviewed by Hannu Konttinen, 1996.

**Vizina, Len** Vice-President and General Manager of CTR Manufacturing Inc. in Union Grove, North Carolina. Interviewed by Ken Drushka, 1996.

**Walbridge, Tom** a retired professor of forest engineering from Virginia Tech; headed the Harvesting Research Project during its existence from 1967 to 1972. Interviewed by Ken Drushka, 1996.

**Watkins, Joe** a retired Georgia Pacific Logging Engineer in Savannah, Georgia. Interviewed by Ken Drushka, 1996.

**Wisotzky, Raul** started at Lokomo and is currently Area Manager for Timberjack's European Distribution Center, based in Tampere, Finland. Interviewed by Hannu Konttinen, 1996.

# Photo Sources

Author's collection 163

Backström, Bo Göran cover, 217

British Columbia Archives and Records Service 31, 42

British Columbia Forest Museum 44, 62

Campbell River Museum Cover, 20, 40, 62

Canadian Pulp and Paper Association 126, 140, 179, 180, 181, 182, 188, 208, 214

Dix, John 163

Enso collection, Lusto 38, 41

Finning Tractor 63, 104

Finnish Forest and Park Service, Lusto 38, 65

Finnish Forest Association, Lusto 58

Finnish Forest Research Institute 8, 14, 32, 33, 37, 41, 55, 57, 60, 61, 66, 68

Forest History Society, US 14, 36, 39, 40, 42, 43, 43, 44, 57, 59, 63, 64, 80, 80, 103, 104, 105, 106, 106, 107, 109, 112, 126, 136

Foster, Harry 179

International Woodworkers of America, Local 1-80 58, 64

Jarck, Walter 88

Lapland forestry museum 34, 39

LeTourneau, Leo 35

Link & Pin Museum, Sayward, BC 56

Malmberg, Carl Einar 88

Metsäkuva-arkisto 114, 128, 134, 142, 161, 192

Ponsse 157, 213

Reid, Bert 37, 81, 162, 181, 189

Sasaki, Shojo 218

Silvatech Industries 214

Skinskattebergs Skogsmästarskolan (forestry school) 67, 87, 130, 131, 135, 138, 151, 164, 185, 187, 191, 207, 210

Skogforsk, picture archieves 67, 68, 82, 83, 90, 130, 131, 153, 188.

'En smedjas förvandling — ÖSAs historia' (book) 82

Skylead Logging Equipment 180

Specialbild AB 156

Studio Mauri Kivistö 186 ja 216.

Swanson, Alf 113

Timbco Hydraulics 215

Timber Harvesting Magazine 35, 81, 84

Timberjack 12, 83, 85, 89, 90, 91, 92, 107, 110, 110, 111, 111, 112, 114, 115, 116, 129, 132, 133, 134, 135, 137, 139, 140, 141, 151, 152, 153, 154, 155, 156, 158, 185, 189, 190, 192, 205, 206, 207, 209, 213, 216, 218

Valmet (Sisu Logging) 211

Vancouver Public Library 66

Virginia Tech, Department of Forestry 86, 91, 92, 105, 108, 109, 127, 127, 129, 137, 138, 157, 159, 159, 160, 162, 182, 183, 183, 184, 209

Walbridge, Tom 86, 108

Williams, Viv 103

# Index


## A

A-frame 51, 93, 96, 103, 117, 150, 168
Aamulehti 48
Abitibi Pulp and Power 146, 168
Adlers, Lars 174
aerial logging 170, 222
Aero Lift 170
Agrip 75
Ahonen Lumber Company 50
air-conditioning 166, 216
Ajax, Ontario 98
Alabama 19, 26
Alaska 21, 27
Alberni Shipyard 196
Alexander Dunbar and Sons 28
Alfta 74, 75, 124, 158, 166
Alfta bands 74
Alimmainen Rautjärvi 32
"alligator" 28
Allis-Chalmers 48, 74
Allison Company 47
Alto, Texas 54
American Civil War 24
American Pulpwood Association 45, 167, 168
American Revolution 22
Anderson, Arne 123
Anderson Woodhawk 77
Andersson, Erland 172
Anglo-Canadian Pulp and Paper 120, 139, 171
Angora, Minnesota 121
Anthony Manufacturing 145
Appalachian Mountains 21, 23, 48, 78
Arbomatik 146, 159
Arbor 52
Archer, Florida 78
articulated steering 77, 103
Ashland Iron Works 52
Athey bummer 66
Athey Truss Wheel Company 48, 145
Atlanta, Georgia 167, 168
Australia 21, 23, 69, 120
Austria 75, 96, 169
axe 29, 57

## B

Baie Comeau 117
Baker, Bob 173
ball-hooting 21, 35
balloon 171
balloons 170, 222
Baltic 22
Baltic Sea 22
Baltimore, Maryland 94
Bamse 74, 89, 90, 122, 123
Baptist, William 27
Barnhart loader 28
Beaumont, Texas 24
beavertail bars 71, 79
Bebo 71
Belamy, Alabama 47
Bell 209
Bell 214 helicopter 171
Bell, Garnet 119
Bell, Irvin 194, 198
Bell, Paul 198
Bellingham, Washington 53
Beloit Corporation 98, 118, 149
Beloit Harvester 118, 123, 136
Berger Engineering 95
Bergslaget 53, 73, 79
Best, Daniel 47
Big Bear 125, 133, 134, 149
Big Stick loader 97, 105
big wheel 24, 36, 48
Billerud 172
Birmingham, Alabama 201
Blacksburg 168
Bloedel, Stewart & Welch 53
Blondin 148, 191
Blonski, Joe 168, 184
Blount Inc. 199
Blount, Wynton 199
Blue Ox 93, 98, 99, 111
BM 400 forest tractor 85
BM-10 82
Bobcat 120, 164
Boeing Vertols 171
Bohemia Lumber 171
Bohemia Lumber Company 99
Bohlinder-Munktell (BM) 73, 74
Bombardier 73, 76, 117, 127, 146
Bombardier HWD 123
Bombardier Processing Unit 117
Bonnard Equipment Limited 77
Bonnard forwarder 98
Bonnard Logger 77, 91
bow guide 54
bow saw 59, 71, 81
Bowater Timber 168
Bowaters 169
Boxer 149
Boyd, Jack 118
Brånemo 165
Brantford, Ontario 119
Brice, Carl 78
Britain 71, 75
British Columbia 20, 21, 31, 40, 42, 44, 46, 50, 53, 56, 62, 64, 66, 103, 104, 145, 146, 162, 163, 168, 169, 170, 171, 175, 179, 180, 214, 219
British Columbia Loggers' Association 53
Brooklyn 50
Brown Machinery Service 97
Brunett 123, 148, 149
Brunswick, Georgia 71, 173
Bruun, Lars 112, 123, 141, 147, 148, 165, 191
Bruun Systems 165, 177, 178
Bruunett 131, 147, 148, 165, 177, 178, 191, 192, 200, 205
Bruunomatic 178
Buchanan, Michigan 77
bucking 51, 76, 120, 173
Buell, Alan 19
Bulldog 50
bulldozer 49, 51
bummer 23, 24, 66
bundle bucking 173, 182
Bunyan, Paul 221
Burch, Tom 16, 17, 96, 172
Burrard Inlet 29
Busch Combine 118, 123, 126, 193, 197
Busch Loader 119, 127
Busch, Tom 72, 118
Butters, Horace 26

## C

cab 216
cable loading systems 51, 85
cable logging 25, 26, 70
cable machines 27
cable yarding 27, 28, 46, 47, 48, 50, 70, 76, 96, 100, 107, 123, 168, 169, 177, 179, 180
Calabari, Nigeria 201
Calgary, Alberta 76
California 25, 27, 47, 49, 52, 171, 175
California sugar pine 24
Cambio 87
Campbell River, British Columbia 95, 171
Can Car 173, 200
Canada 22, 23, 31
Canadian Hornet 71

Canadian International Paper (CIP) 98, 119, 146, 167, 168
Canadian Pulp and Paper Association (CPPA) 45, 51, 76, 77, 98, 118, 167
Carco winch 144
Carroll, Joseph 24
Catawba Timber 79
Caterpillar 47, 48, 63, 70, 99, 104, 117, 120, 146, 193, 195, 201
Caterpillar Tractor Company 24, 47
Cavalier Auto Service 145
Cemet-Agrip 201
Centipede 30
Central Forest Association Tapio 124, 149
chain flail delimber 117, 172
chain saw 51–54, 56, 58, 70, 72, 76, 100, 101, 117, 121, 147, 166, 172, 193
Challenger, Dave 79
Challenger, Jack 53
Chicago 54, 71
chienning 22
Chile 203
China 21, 150
Chip'N Saw 119
chipping 148, 171, 173, 197
clambunks 121, 146
Clark Equipment 77, 145
clear-cut logging 26, 177, 221
Cletrac 74
Cletrac crawler tractor 41, 50, 167
Cleveland, Ohio 50, 146, 200
Cliffe, Joe 150
Climax 25
Clova, Quebec 168
Clowez 75
Clyde Iron Works 26, 28, 46
Clyde skidder 43
Coca Cola 78
cold decking 27, 46
Columbia 181
Columbia, South Carolina 79
Combi 188
Combine 119
Comet 71
Companion 71
computerized measuring and control system 216
computers 18
Connecticut 52
Consolidated Paper 171
container system 161
control system 216, 217, 222
conversion of forest land 220
Courtenay, British Columbia 169
Cox, Joe 71
Cranab 125
Crawford, Mal 18, 95, 121, 166
Crawford, Pat 94, 194, 195, 215
crawler tractor 38, 41, 47–49, 54, 68, 70, 72, 74, 95, 98, 146
Crown Zellerbach 100, 170
Cryder, Morris "Doc" 144, 147
CTR 194, 197, 199
Currie Cost Cutter 122, 173, 182
cut-to-length 16, 114, 122, 123, 128, 135, 147, 149, 165, 166, 199, 201–204, 222
cutter 60
Cyclo-Crane 170
Cypress 179
cypress swamp 27, 28, 53
Czechoslovakia 117, 177

## D

Danarm 71
debarking 60, 61, 87, 115, 146, 148, 150, 172, 173, 176, 197
DeBeck, E.K. (Ned) 29
delimbing 117, 119, 120, 130, 146, 148, 165, 172, 175, 196
delimbing chains 155
delimbing gate 172
delimbing machine 75
Denharco 200
Denis and Harricana 200
developed countries 220
developing countries 220
diesel motor 42, 46, 50, 70, 71, 93–96, 111, 124, 145, 149, 169, 171, 175, 198, 204, 222
Dinos crawler tractor 38
Disston 76
Dixie Pallet System 78
Dolbeer, John 25, 204
Dolbeer steam donkey 25, 26, 42, 204
Dolmar 52, 53, 71
Dolmar Machinfabrik 52
Domän 113, 123, 125, 165
Domtar 118
Donaren 185
donkey 25–27, 123
Doroverken 177
Douglas fir 21, 44
Dowty Equipment 98
Dowty forwarder 98, 123
drag saw 53
Drake, Roger 144, 145
Drivax 125, 131, 148, 149
Drott 196
Drott, Erv 121, 194
Drott feller buncher 121, 137, 140, 173, 193
Duluth, Minnesota 26, 145
Duplex 62
"duplex" loading 27
Durban 174

## E

East Thurlow Island 31
Eaton Corporation 200
Eau Claire, Wisconsin 30
Ecologger 169, 180
Egypt 17
Eitel, George 22
electric generator 46
Electric Logger 93
electric saw 51, 52
elephant logging 37
Ellicott Corporation 94, 200
Ellingsons 21
Elmia Exhibition 148, 169
Elofson, Hans-Åke 174, 197
Ely, Minnesota 98, 118
Empire Manufacturing 26
Enderby, British Columbia 169
Enso 175
Enumclaw, Washington 95, 200
environmental movement 175, 220
Ergo Series 213
Erickson, Jack 171
Erie, Pennsylvania 25
Escanaba, Michigan 145
ESCO 146
Eskelinen, Väinö 65
Eskisltuna 73
Espanola, Ontario 93
Estes, Harold 194
Eureka 25
Evo Forestry School 57

## F

falling 51
farm tractors 82, 165
Fedje, Olaf 51
feed rollers 115, 155
feller buncher 121, 146, 166, 168, 172, 177, 193, 194–196, 199, 212
feller forwarder 208
feller skidder 117, 120, 127, 146, 148, 159, 165, 176
felling head 134, 155
Filipstad 165
Finko 186
Finland 25, 30, 46, 71, 72, 74, 76, 102–124, 148–150, 169, 171, 172, 175, 177, 178, 199–201, 203, 204
Finnish Natural Resources Research Institute 148
Firestone 144
Florida 16, 69, 72, 78, 100, 144, 172
Flynn, Tom 70
FMG Timberjack 201
Food Machinery Corporation (FMC) 146, 162
Ford 70, 144
Ford Brunett 149
Ford tractor 123

Fordson 46, 74
fore-and-aft road 49, 50, 64
Forest Ant 164, 177
forest ecology 221
Forest Engineering Research Institute of Canada 196
forest landscape 222
Forest Machine Group (FMG) 201
Forest Owners' Association 175
Forestry and Forest Products Research Institute 222, 204
Forestville, Quebec 171
Forrex 200
Fort Bragg, California 44
Fort William, Ontario 121, 200
forwarder 73, 75–77, 79, 98, 119, 123, 125, 145, 148, 165, 166, 176, 223
Fourmies, France 75
frame steering 79, 91, 92, 99, 107, 112, 123, 125, 149, 158, 177
France 77, 177
Frank Hough Company 145
Franklin Equipment 145, 194, 200
Franklin, Gordon 172, 183
Franklin Loggers 121, 145
Franklin Prehauler 157
Franklin River camp 53
Franklin skidder 167
Franklin, Virginia 145
Fraser, Frank 46
Fraser River 29
full-tree logging 114 117, 123, 128, 135, 149, 150, 165, 193, 201, 203, 204, 222
FWD Truck Company 93


## G

G.B. Murdie Company 46
Gafner 194
Gafner, Emil 145, 202
Gafner Iron Mule 145, 159, 202
Garnet Bell 204
Garp grapple 138, 147, 148
Garpenberg Forestry College 73, 147
Garrett, Dwight 95, 122, 200
Garrett Tree Farmer 99, 109
gas donkey 46
gasoline 46
Gaspe 168
Gearmatic winch 47
General Motors 70
Georgia 71, 80, 86, 168, 173, 175, 194
Georgia Pacific 100, 196
Gerber 52
Germany 52, 53, 54, 70, 71, 166
Gilchrists 21
Go Getter 78, 86
go-devil 23
Go-Tract Systems 76
Godans 148
Goodyear 50, 144, 145
Grabinski, Phil 96, 104
grader 88
Grannäs 73
grapple skidder 120, 145, 146, 172, 173, 193, 194
grapple yarder 99, 170, 179, 196
Great Depression 219
Great Dismal Swamp 145
Great Lakes 26
Great Lakes Forest Industries 121
Greece 21
Green Swamp 26
Gulf Engineering 145
Gustafsson, Gösta 170
gyppo logger 71


## H

Hagglund & Soner 125
Hahn Harvesters 175
Hahn, Ray 97, 173
half-track 82
Halmstad 178
Hamburg 52
Hamilton, Doug 146
hand jack 21
handfaller 56
handlogging 21, 22, 29, 168
Hankkija Co-op 124, 150, 175
Harambourne, Pete 49
Härmälä 201
Harricana 196
Harrison Pulpwood Harvester 79, 86, 123
harvester 117–123, 154, 166, 175, 176
harvesting head 186
Harvesting Research Project 167, 168
harvesting system 156
Hasab 199
haul roads 23
hauler 39
Havel, Vaclav 223
Hawker Siddeley 200
Hawthorne, Florida 144
Hayes 70, 100
Hayes-Anderson Motor Company 50
Hayes-Lawrence 50
hayrack boom 105
Hede Mekaniska 135
Hedegård, Bo 73, 79, 92
heel booming 27, 28
Heikkinen, Leo 94, 98, 106, 194
Heikkinen Machine Company 199
Heisler 25, 40
helicopter 171, 222
Helin, Esko T. 25
Helmer, Sten 122
Helsinki 196
Hemek 199
Hemek All-driv 135, 148
Heron Bay, Ontario 117
Hiab 79, 92, 98, 149
Hiab "Forest Elephant" loader 112, 123, 125
Hiabob 98, 118
high lead logging 19, 26
Hingman, John 167
Hinteregger 169
Hinton, Alberta 117
Hoffman, Bill 144
Holstrom, John 50
Holt brothers 47
Holt crawler 66
Homelite 71, 76
Horncastle, Clark 172
Hornet power saw company 54
horse 22, 23, 28, 48, 51, 55, 76, 103, 123, 124, 146, 165
horse logger 101
Hovercraft 221
Hudson Sales Corporation 201
Husqvarna 71, 188
Huusko, Pertti 196
hydraulic knuckle boom 91
hydraulic loader 87, 94, 96, 158
hydraulic system 219
hydraulics 79, 143
Hydro-Ax 183, 193, 194, 196, 199, 214
hydrostatically driven skidder 214
Hyster 48, 63, 76


## I

ice road 50
Idaho 52, 102
IEL chain saw 58
Iglund-Jones Mini-Alp 169
Iisalmi 38, 176
imbuyah 27
Indonesia 203
Industrial Engineering Limited (IEL) 54, 71
Industrial Revolution 17
International Forest Products 171
International Harvester 48, 145
International Paper Company 72, 120, 167
International Timber 49
Iron Mule 145, 159, 202
Ironwood, Michigan 50
Italy 177
Iwafuji 177


## J

J.I. Case Machine Company 121
jack 29
Jämland 123
Japan 17, 23, 177
Japan's Forestry and Forest Products Research Institute 218

Jarck Go-Getter 78, 86
Jarck, Walter 79, 86
Jari, Brazil 189
Jesus Machine 120
jinkers 23
Jo-Bu 71
Joensuu 175
John Deere 145, 149, 193, 194
Johnson, Bill 49
Jonsered 165

## K

Kaarnametsä, Matti 167
Kaifeng 21
Kajaanin Automatiikka 199
Kämpe, Lars-Göran 197
Kare, Erkki 200
Karl Gustaf, King of Sweden 169
Karlsson, Börje 148
Kelly, Tom 70, 73, 122
Kelowna Machine Works 145
Kemi Oy 25, 30
Kennebec 28
Kenworth 50, 70, 100, 104, 145
Kerr, Archibald 49, 93
KF cooperative 74
Khrushchev, Nikita 125
Kimberly-Clark 168, 175
King, John Leslie 26
Kinston, North Carolina 145
Kiso Forest 21
Kita 71
Knight wheel saw 80
knuckleboom 91, 118, 121, 165
knuckleboom loader 18, 98, 102, 175, 197
Kockums 125, 130, 141, 147, 149, 151, 156, 165, 166, 173, 177, 187, 200
Kockum 826 forwarder 148
Kockum Bruunett 148, 165
Koehring Shortwood Harvester 126
Koehring-Waterous 119, 193, 195, 196, 201, 208
Koilliskaira 25, 30
Koirajärvi 72
Koller 169
Komatsu 201
Kootenay Equipment 146, 162
Kopparfors AB 73, 125
Koroleff, A.M. 45
kuda-kuda 22
Kuolajärvi 65
Kurelek, John 193, 195, 196
Kut-Kwick 71, 80
KVP 93

## L

labour conditions 22, 52, 69
labour-research institute (SDA) 102
Lachine, Quebec 77
Lagerlöf 38, 46
Lagus, Mauri R. 60
Laine, Muisto 164, 177
Lanford, Bob 221
Lang, Alfonso 52
Lapland 30, 34, 149, 177
Lapsley, James V. 196
Larson, Bob 98, 118, 121, 136
Larsson, Calle 73, 79
Larsson, Lasse 166
laser technology 221
Lathrop 30
Laurel, Mississippi 24
Lawrence 10–10 donkey 47
Lawrence 36 yarder 47
Lawrence, Frank 47, 50, 76, 96, 170
Lebanon 16, 17, 21
Lerp, Emil 52
LeTourneau, Bob 49, 93, 100, 103, 144
LeTourneau Log Transporter 100
LeTourneau's bulldozer 50
levelling cab 152
Liberia 81
Libertyville, Illinois 145
Lidgerwood Manufacturing 26, 27, 46, 96
Lidgerwood Portable Logger 27, 47
Lima Locomotive Works 25
limbing gate 183
Lindsey, John 24
Link Brothers Welding and Machine Shop 72
Linn 50
Little Bear 125, 132, 149
Little Skotten 747 177
Little Tugger 46
loader 17, 18, 27, 28, 36, 40, 43, 51, 68, 73, 79, 85, 87, 90, 94–102, 104–106, 109, 112, 116, 119, 120, 122–125, 127, 132, 148–150, 158, 166, 168, 170, 173, 175, 182, 194, 196–199, 201, 209, 217, 219, 223
loading grapple 91
loblolly pine 19
locomotive 24, 25, 27, 40
Log All 146, 159
log loading 51
log skidding 48
log sled 83
Loggermobile 94, 107
Logger's Dream 96, 98, 101, 102, 108, 145, 167, 168
logging camp 34
logging railroad 101
Logging Research Associates 146, 159, 167
logging sulky 48
Logging Tractor 146
logging truck 50, 72
Logma 147, 151
Lokomo 129, 149, 154, 167, 175, 178, 200
Lokomo 990 harvester 197
Lokomo 990/910 209
Lokomo Forest Oy 201
Lombard, Alvin 28, 47, 50
Lombard Log Hauler 39, 42
Lombard steam-engines 25
London, Jack 25
Long Lisa forwarder 125
Longlac, Ontario 168
Longlac Pulp and Paper 19
longleaf pine 28
Los Angeles 79
Lotta 79, 92
Louisiana 118
Louisville, Mississippi 96, 97, 101
Lowther wheel saw 71, 80
Luban 52
Lundahl, Robert 123
Lusaka 174
Luthman, Gösta 45

## M

MacDonald Steel 195
Mack 100
Mack brothers 50
Mackmyran stump extractor 178
MacLeod-Smith, Ian 201
MacMillan Bloedel 99, 170, 175
Madill 196
Madill, Chuck 95
Madill steel spar 104
Magill, Wes 200
Mahood, Ian 171
Maine 28
Makeri 164, 177
Mäki 169
Malcolm Island, British Columbia 64
Mall Tool Company 54, 71
Malmberg, Carl Einar 75, 88
Malmö 125
Mammoth Tree Shear Company 72
Manistee, Michigan 24
Manitoba 183
Manitoba Forestry Resources Ltd. 172
Marathon 72
Marathon Paper Mills 118, 168
Marion 28
Mark II Forwarder 77
Marttiini, Lauri 149, 176, 177
Massachusetts 52
Massey Ferguson 140, 145, 172
Master Speed skidders 145

Matheson, Chester 170
McColl, Bruce 98
McCulloch 70, 71, 79, 219
McGiffert, J.R. 26, 36, 43, 96
McGiffert loader 28
McGill, Wes 76, 94
McLean, Claude 27, 44
mechanical saw 53
Mekaniska Verkstad AB 125
Menzie Muck 204
Metsäpalvelu Oy 150
Metsäteho 175
MF-Kärppä 124
Michigan 22, 24, 26, 36, 50, 70, 74, 77, 117, 145, 173, 193, 202
Michigan PL-75 77
Michigan State University 76
Mickelsson, Stina 166
Middlesex 75
Miettinen, Julius 30
Miller, Spencer 26
Mini Bruunett 177, 178, 191, 192, 200, 205
Mini Skidder 145
Mini-Alp 179
Minneapolis, Minnesota 145
Minnesota 22, 26, 28, 30, 98, 118, 121, 136, 145, 173, 175
Mississippi 24, 26, 96, 97, 101, 108, 112
Mobile 70, 73
mobile loader 28
mobile spar yarder 95
MoDo Mekan 125, 148
Moisander, Jukka 169
Montague slasher 72
Montana 30, 47, 72, 182
Montreal 23, 170
Mor-Bell faller 209
Morbark 173, 194, 197, 198, 209
Morenius 165
Moresby Island 175
Morey, Norval 76, 173, 176, 194
Morgan SilvaCom 214
Mosher, Bill 168
Mosher, Frank 171
motor truck 47, 49, 54
motorized sled 38
Motormule 75
MS-Robur 125
mule 23, 76–78, 94
Myanmar 37

## N

Nanaimo 95
Napco Crab 145
Napco Industries 145
Napoleon 22
narrow-gauge 24
National Board of Forestry 149
natural forest 223
Natural Resources Research Institute 150
Nelson frame 76
Nesco Slashmobile 72
NESCO Woodsmobile 100 Harvester 138
New Brunswick 22, 28, 126, 169
New England 22
New Hampshire 24
New Orleans 27
New Westminster, BC 26
New York 22, 26, 51
New Zealand 23, 69
Newfoundland 169
Nile Valley 17
Nitnat 170
Nodwell 76
Norcar 192, 199
Norie, Frank 71
Norsälven Sågverk AB 122
North Carolina 26, 35, 47, 136, 145
North Shore Paper Company 117, 167
North Western Pulp & Power 117
Northern Engineering and Supply Company 121
Norway 71, 169, 177
Nova Scotia 22, 168

## O

Oakdale, Louisiana 43
Ocala, Florida 72
Ohio 25
Okanagan Helicopters 171
Okefenokee Swamp 28
Oksa-Olli 75
Oliver Corporation 74
Olla, Louisiana 72
Olsson-Lokind, Per O. 45
Olympic Peninsula 50
Omark Industries 98, 121, 166, 199
Ontario 22, 27, 28, 45, 54, 69, 72, 76, 91, 93, 94, 96, 98, 110, 118, 119, 121, 139, 144, 168, 172, 175, 195, 196, 200, 208
Ontario Paper Company 117, 172
Opel 65
Oregon 49, 52, 93, 96, 99, 105, 145, 146, 169, 170, 171, 173, 194, 199, 201, 202
Oregon Saw Chain 71
Ösa 73, 74, 85, 89, 125, 132, 148, 152, 153, 155, 156, 158, 165, 166, 167, 175, 177, 185, 189, 190, 200
Ottawa River Valley 22
Oulu 74
Overpack, Sylas 24
Owren Yarding Systems 169
oxen 23, 31

## P

Pacific 100
Pacific Car and Foundry 120
Pacific Logging Congress 53
Packman Machinery 75
pallet cart 92
pallet 78
Papunen, Kauko 145
Parson's Whittemore & Lyddon 201
Partington, Bob 147
Partner 71
Patterson, Ken 120
Paylogger skidder 145
Peerless Corporation 201
Peninsula Iron Works 52
Pennsylvania 25, 28
Performer 93
Perunkajärvi 46
Peterbilt 100
Peterson, Howard 50, 52
Peterson, Nels 173
Peterson Pacific 173, 197
Pettibone 145, 193
Phoenix 39
Phoenix Manufacturing Company 25, 30
Pierce-Arrow passenger car 50
Piironen, Jaakko 102, 150
Pika 50 processor 142, 150
Pika 75 harvester 151
pine forest 71, 77, 118
Pino-Teho 150
Pinomäki, Sakari 142, 148, 150, 176
Pioneer bow saw 59
Pioneer Timber Company 64
plank road 44, 50
plantation 17, 221, 223
planting 206
planting machine 134, 223
Plumett 75
Plustech 201, 203, 204, 218
Poland 52, 177
Pond, Leon 19, 99
Ponsse 178, 199, 213
Ponsse PAZ 157, 176
Pope, John 117
Porta-Tower 95
portable skidder 27
portable skidders 27
portable steel spar 95
Portland, Oregon 26, 52, 71, 76, 93, 145
Potlatch Lumber Company 52
Poulan, Claude 54, 71
power saws 70, 72, 96
prehaul skidders 47
Prentice loader 18, 95, 106, 199
Prentice, Reed 54
Prentice Saw Buck 106, 173
Prentice, Wisconsin 98

processor 117, 121, 145, 146, 147, 148, 149, 153, 154, 155, 165–167, 169–171, 173, 175, 176, 177, 178
pull-boats 27
pulp 17, 19, 45, 51, 59, 60, 69, 76, 80, 84, 86, 101, 107, 120, 139, 165, 16–169, 171, 172, 173, 175, 194, 195, 197, 198, 199, 202
pulp mill 69
Pulpjack 160
pulpwood 45, 51, 72, 74, 76, 78, 97, 98, 117, 119, 126, 167, 171, 173, 177
Pulpwood Dream 98

## Q

Qin Feng 21
Quebec 22, 28, 45, 69, 76, 77, 78, 96, 117, 127, 129, 146, 159, 167, 168, 169, 171, 181, 196, 200
Quebec's St. Lawrence River 22
Queen Charlotte Islands 93
Queen Elizabeth 171

## R

R.L. Windsor and Sons 120
Raahe Timber Products Company 46
Radisson Tractor 76
railway logging 54, 219
railways 24, 48
Raket 71
Ralston, Ken 167
Rauma-Raahe 74
Rauma-Repola 175, 200, 201
Redonda Island 171
redwoods 175
Reed Prentice Company 52
Reed Prentice saw 70
reforestation 124, 134
Reforester 163
regeneration harvesting 211
reindeer 30, 34
Rhode Island 168
Rhodesia 174
Rinco 52, 58
Rio Negro River 27
river driving 23, 28
road building 50
Road Machines Limited 75
road monkey 22
roader 25
roader donkey 44
Robbinsville, North Carolina 35
robotic logging machines 223
Robur 124
Robur Flexor 148
Roman empire 21
Rosedale Equipment 169
Rosenberg and Kruuse 178
Rottne 148, 178, 188, 191, 199
Rovaniemi 30, 46, 149, 176, 177
Royal Navy 22
rubber tires 143
rubber-tire loader 99
rubber-tired skidder 93–95, 99, 107, 143
Ruovesi 48
Russia 30, 146, 177, 196
Rysä, Mikko 200

## S

San Antonio, Texas 72
Sandberg brothers 125, 131
Sandberg, Hugo 25, 30
Sarawack 22
Saskatchewan 30
Savannah 194, 198
Sawbuck 166
scarifying 124, 148, 185
Scotland 169
Seattle 26, 50, 95, 120, 122
Second World War 74
second-growth pine forest 69
Sector 52
Sedro Woolley 46
Serla 161
Serlachius 161
Shawano, Wisconsin 195
Shay, Ephraim 24, 40
Short Dog 145
Short Lisa skidder 125
short-wood logging 76
shovel loader 99
SHS planting machine 134
Siberia 203
Sicard 120
Siiro, Nestor 121
Siiro-Omark harvester 138
Sikorsky 171
Silva Nova 206
Silver Grizzly Logging 171
silviculture 17, 18
Simcoe, Ontario 28
Simula, Markku 220
simulator 203
single-grip harvester 121, 138, 178, 186, 187, 199, 202, 211, 213
Sipilä, Olavi 124
Sisu 149, 202
Sitka spruce 21, 56
Skagit 46, 70, 95, 96, 99, 100, 170
Skagit Little Tugger 62
skid road 25
skidder 18, 19, 43, 77, 94, 97, 99–103, 117, 122, 124, 125, 128, 144, 149, 169, 170, 175, 177, 194, 200, 212
skidding 23, 75, 93
Skogs-Basen 185
Skogsarbeten 165, 178, 193
Skogsjan, 178, 187
Skruven 130, 147
Skycar 96
Skyhook Enterprises 96, 104, 171
Skylead Logging 169, 180
skyline system 26
skyline 96
slasher 72, 106, 166, 173, 175, 193, 196, 197, 199
sled 22, 23, 28, 34, 35, 37, 38, 41, 42, 62, 68, 76, 83, 89, 95, 96, 168
sleigh hauling 22, 23
sling loader 73
slip tongue 24
sloop 76
Small, Don 124
Smith, D.H. 54
Smith Timbermaster 169
SMV 149
snakers 27
snigging 23
Snobenek, Mark 72
Snoken 210
Snow Comet 73
snowmobile 73
soil preparation 90
Somater 56 75
Sonora Island 100
South Africa 169, 174, 194, 198, 209
South Carolina 47, 79, 86
Southern Brazil Lumber Company 27
Soviet Union 53, 69, 71, 72, 75, 101, 125, 147
SP 21 felling head 206, 207
Spain 164, 177
Sparmatic 96
spar 22, 95
Spokane 52, 195
Springhill, Louisiana 145
Sproat Lake 170
Spruce Falls Harvester 140, 146
Spruce Falls Power and Paper Company 121
squared timber 23
Sri Lanka 162
steam donkey 26, 42, 46, 95
steam engine 17
steam hauler 28, 30, 48
steam loader 36, 40
steam tractor 47
steam yarder 26
steel spar 182
steel-wheeled arch 63
Stensele Mekaniska Verkstad 125, 131, 166, 199
Stephenson, Ev 194, 195, 198
Stevenson, Russell 198
Stewart, Faye 171
STFI 138
Stihl 53, 54, 58, 70, 71
Stihl, Andreas 53
Stillwater 66
Stockton Wheel Company 24

Stora Kopparberg 53
Storebo Bruk 148
Straussler Lypsoid 75
stroke delimber 172, 181
Stuart, Bill 77, 78, 102, 118
stump extractor 166, 189
Sumner, Washington 63
Sunds 125
Sunds debarker 115
Sunds processing system 116
Suojärvi 41
sustained-yield forestry 221
Swamp Buggy 145
swamp-logging 27
Swanson, Alf 113
Sweden 45, 50, 71–74, 79, 94, 102–125, 146–150, 165, 169, 170, 172, 174, 177, 178, 193, 199
Swedish Mechanization Foundation 123
swing yarder 168
Switzerland 96, 177, 204
Sylva 52
Symons, Bob 76, 94, 144, 168

## T

tail processor 210
Talladega, Alabama 200
Tampere 149
Tanguay, Jean-Paul 196
Tasmania 21
Taylor Machine Works 96, 98, 101, 108, 112, 145
Taylor Prehaul Skidder 100
teamster 24
Tehdaspuu 171, 175
Telecarrier 169
Teli-Lokkeri forwarder 150
Tennessee 18, 168
Terra 865 BK 149
Terra AK 149
Texas 16, 24, 27, 28, 54, 69, 72
The Pas 172
Thibodeau, John 78
thinning 9, 103, 119, 148, 164, 165, 169, 177, 178, 200, 202, 211, 223
Thomson, T.M. 171
Thunder Bay, Ontario 208
Tidewater Equipment 71, 78
Tidewater Manufacturing 173
Tigercat Industries 195
Timbco 215
Timbco Hydraulics 195
Timber Tosser 97, 98, 109
Timber Toter 145
Timberbuncher 118
Timberjack 18, 94, 95, 97, 99, 119–121, 123–125, 138, 144, 146, 149, 160, 163, 167, 168, 193, 195, 200, 201
Timberjack 1270B 216
Timberjack 200 skidder 122
Timberjack 460 skidder 212
Timberjack 570 harvester 213
Timberjack 600 feller buncher 212
Timberjack 810B forwarder 217
Timberjack RW-30 harvester 137
Timberjack Shortwood Harvester 139
Timberjack skidder 111, 113, 124, 162, 189
Timberland Machines 76, 94, 96, 97, 107, 118, 168, 200
Timberline Equipment 194
Timberskidder 94, 110, 124, 200
Timmer-Kalle 125, 133, 149
tire 50, 144, 147
Titan 71
TLD Gauthier 169
Toivakka 171
Tom Cat 70
Tomahawk Timber 117
tongs 24, 67
Tornado 71
Tote-A-Load 112, 145
Tournarch 93, 103
Tournaskidder 93
tracked skidder 146, 162
tractor 70, 72, 73, 75, 77–79, 93, 99, 101, 124, 143, 146, 177, 199
trailer 89
tramway 23
Tree Farmer 95, 97, 122–124, 125, 200
tree shear 72
tree-planting 148, 163
Treever 140
Tres Barras 27
Tri-Track Mover 204, 218, 222
tropical hardwood 189
truck logging 41, 49, 51, 69–71, 78, 100, 102
Truro, Nova Scotia 145
Tume-Tapio 124
Turku 52
Two Harbors, Minnesota 175
Typhoon 71

## U

Union Camp 145, 194, 199
United 144
Uppland 53
Urus yarders 169
US Forest Service 70

## V

Valmet 71, 114, 124, 148, 149, 158, 175, 176, 178, 200, 202, 207, 210, 211
Vancouver, British Columbia 26, 50, 100, 54
Vancouver Island 58, 170
Vancouver Island Helicopters 171
Vargön 172
Värmlands Skogsarbetsstudier (VSA) 45, 141, 148, 172
Victoria, British Columbia 46
Victoria Falls 174
Vidgrén, Einari 157, 176, 178, 199
Vienna 52
Vindelbjörn 125
Virginia 27, 121, 145, 157
Virginia Polytechnic Institute 168
Vit Feller Buncher 127
Vit Feller Skidder 123
Vit Forwarder 129
Vit, Rudy 117, 120, 127, 166, 196
Vitak and Rainer Company 52
Vizina, Len 193, 194, 202, 203, 214
Vollmershausen, Dennis 200
Volmari 176
Volvo 73, 74, 125, 132, 148, 149, 165, 170, 174, 176, 177, 178
Volvo 868 forwarder 156, 158
Volvo BM 79, 148, 165
von Heideken, Folke 123
VSA forest tractor 112

## W

Wagner Tractor 94, 107
Wainwright, Bill 170
Walbridge, Tom 72, 74, 167, 168, 171
Wales 51
Walking Machine 204, 218, 219, 222
Wallace and Harper Baikie 95
Warner and Swasey 146
warping tug 28
Washington 14, 46, 49, 50, 53, 63, 95, 99, 103, 170, 200
Washington Flyer 47
Washington Iron Works 26, 171
Waterous, Charles 119
Watkins, Joe 47, 53, 100, 101
Wausau, Wisconsin 121
Weasel 181
West, John 28
Western Lumber Company 30
Westfall, Paul 93
Westminster Iron Works 26
Westvaco 184
Weyerhaeuser 103
wheel saw 71, 80, 167
wheeled arch 48
Whirlybird 145
White 64, 100
White fingers syndrome 147
White Motor Company 50, 74
White Owl Parts Company 145
white pine 22
Willamette Iron & Steel Works 26
Williams, Norman 169

Williams, Viv 93
winch skidding 114
Windsor Tree Harvester 120, 137
Winn, Michigan 173
wire rope 24, 46
Wisconsin 22, 28, 30, 93, 98, 106, 121, 194, 195, 215
Wisotzky, Raul 201
Wolf, Charles 52
Wolverine 194
wood-gas generators 72
Woods, Bob 100, 169
Woodsmobile 121
Woodstock 28, 76, 97
Work Efficiency Association 150
World Forest Congress 122
World War Two 63, 69, 77, 219
Wyssen 96

## Y

Yaak River 182

## Z

Zambia 174

# Global Forest Resources – 1994

Source: F.A.O. Forest Products Yearbook 1994
Forest Resources Assessment 1990
Symbols: CUM = Cubic Metres, F = F.A.O. Estimate

| Country | Land Area 1 000 ha | Forest Area 1 000 ha | Roundwood production 1 000 CUM | Roundwood Imports 1 000 CUM | Roundwood Exports 1 000 CUM | Fuelwood+ Charcoal production 1 000 CUM |
|---|---|---|---|---|---|---|
| **Developed countries** | | | | | | |
| Finland | 30 462 | 20 112 | 47 928 | 7 563 | 1 700 | 4 101 |
| Sweden | 40 823 | 24 437 | 66 024 | 7 524 | 1 271 | 3 824 |
| Norway | 30 688 | 8 697 | 8 744 | 3 414 | 540 | 470 |
| **Nordics** | **101 973** | **53 246** | | | | |
| Belgium-Luxemburg | 3 282 | 705 | 4 340 | 4 087 | 1 690 | 550F |
| Denmark | 4 253 | 466 | 2 361 | 604 | 389 | 491 |
| Iceland | 8 844 | 0 | | | | |
| Ireland | 6 890 | 396 | 2 636 | 76 | 424 | 60 |
| Netherlands | 3 392 | 334 | 1 999 | 852 | 841 | 210 |
| United Kingdom | 24 086 | 2 207 | 11 285 | 595 | 59 | 260 |
| **North Western Europe** | **50 490** | **4 023** | | | | |
| Austria | 8 254 | 3 877 | 17 692 | 8 146 | 2 179 | 3 259 |
| Former Czechoslovakia | 12 540 | 4 491 | 19 028 | 228 | 2 723 | 1 534 |
| France | 54 325 | 13 110 | 49 617 | 2 917 | 3 272 | 10 454F |
| Germany | 34 934 | 10 490 | 37 012 | 2 633 | 7 121 | 3 795F |
| Poland | 30 445 | 8 672 | 18 544 | 6 | 1 011 | 1 923 |
| Switzerland | 3 976 | 1 130 | 4 974 | 1 144 | 1 238 | 1 153 |
| **Central Eastern Europe** | **144 474** | **41 770** | | | | |
| Albania | 2 775 | 1 046 | 409 | | 1 | 346 |
| Bulgaria | 11 019 | 3 386 | 3 565F | 3 | 26 | 1 728F |
| Cyprus | 916 | 140 | | | | |
| Greece | 12 926 | 2 512 | 2 779F | 39 | 59 | 1 519F |
| Hungary | 9 213 | 1 675 | 4 846 | 168 | 662 | 2 150 |
| Israel | 2 031 | 102 | | | | |
| Italy | 30 128 | 6 750 | 10 031 | 7 519 | 15 | 5 547 |
| Romania | 22 954 | 6 190 | 13 203 | 27 | 38 | 2 963 |
| Turkey | 77 079 | 8 856 | | | | |
| Former Yugoslavia,SFR | 25 540 | 8 371 | 5 269 | 377 | 632 | 1 430 |
| **South East Europe** | **194 581** | **39 028** | | | | |
| Portugal | 8 655 | 2 755 | 11 293 | 1 113 | 714 | 598F |
| Spain | 49 937 | 8 388 | 15 981 | 1 354 | 283 | 2 728 |
| **Iberia** | **58 592** | **11 143** | | | | |
| **Europe** | **550 367** | **149 295** | **359 559** | **50 410** | **26 889** | **51 092** |
| **Former USSR** | **2 138 999** | **754 958** | **183 543** | **150** | **19 916** | **30 800** |
| Canada | 921 500 | 247 164 | 187 951F | 6 269 | 2 677 | 6 897F |
| USA | 913 658 | 209 573 | 491 748 | 2 141 | 20 001 | 92 023 |
| **North America** | **1 835 158** | **456 737** | **751 835** | **8 608** | **22 992** | **160 948** |
| Australia | 754 402 | 39 837 | 21 560 | 14 | 7 341 | 2 898F |
| Japan | 36 460 | 24 158 | 32 722 | 45 887 | 21 | 360F |
| New Zealand | 26 784 | 7 472 | 16 833 | 3 | 5 513 | 50F |
| **Developed Asia / Oceania** | **817 646** | **71 467** | **71 115** | **45 904** | **12 875** | **3 308** |
| **Total developed countries** | 5 342 170 | 1 432 457 | | | | |

| Country | Land Area 1 000 ha | Forest Area 1 000 ha | Roundwood production 1 000 CUM | Roundwood Imports 1 000 CUM | Roundwood Exports 1 000 CUM | Fuelwood+ Charcoal production 1 000 CUM |
|---|---|---|---|---|---|---|
| **Africa** | | | | | | |
| Burkina Faso | 27 380 | 4 436 | 9 771F | | | 9 330F |
| Cape Verde | 403 | 16 | | 1F | | |
| Chad | 125 920 | 11 438 | 4 406F | | | |
| Gambia | 1 000 | 98 | 1 209 | | | 1 096 |
| Guinea-Bissau | 2 812 | 2 022 | 577F | | 5 | 422F |
| Mali | 122 019 | 12 158 | 6 340F | | | 5 936F |
| Mauritania | 102 522 | 556 | 13F | | | 8F |
| Niger | 126 670 | 2 562 | 5 671F | | | 5 321F |
| Senegal | 19 253 | 7 656 | 5 105F | 67 | | 4 404F |
| **West Sahelian Africa** | **527 979** | **40 941** | | | | |
| Djibouti | 2 318 | 22 | | | | |
| Ethiopia,former PDR | 110 100 | 14 354 | 46 980 | | 3 | 45 254 |
| Kenya | 56 969 | 1 305 | 40 372F | | | 38 480F |
| Somalia | 62 734 | 758 | 8 634F | | | 8 529F |
| Sudan | 237 600 | 43 179 | 24 742F | 1F | | 22 453F |
| Uganda | 19 955 | 6 366 | 16 682F | | | 14 464F |
| **East Sahelian Africa** | **489 676** | **65 983** | | | | |
| Benin | 11 062 | 4 961 | 5 726F | | 2 | 5 414F |
| Côte d'Ivoire | 31 800 | 10 967 | 14 487 | | 437 | 11 202F |
| Ghana | 23 002 | 9 608 | 27 140 | | 423 | 25 190 |
| Guinea | 24 586 | 6 696 | | | 28 | 4 296 |
| Liberia | 9 675 | 4 639 | 6 183F | | 600 | 5 202F |
| Nigeria | 91 077 | 15 785 | 108 059F | 2 | 72 | 99 796F |
| Sierra Leone | 7 162 | 1 895 | 3 247F | | | 3 123F |
| Togo | 5 439 | 1 370 | 1 326F | | | 1 122F |
| **West Africa** | **203 803** | **55 919** | | | | |
| Cameroon | 46 540 | 20 366 | 15 040F | | 1 116 | 12 062F |
| Central African Republic | 62 298 | 30 568 | 3 762 | | 37 | 3 250F |
| Congo | 34 150 | 19 902 | 3 639F | | 588 | 2 288F |
| Equatorial Guinea | 2 805 | 1 829 | 714 | | 217 | 447F |
| Gabon | 25 767 | 18 256 | 4 445F | | 1 331 | 2 812F |
| Zaire | 226 760 | 113 317 | 45 927F | | 79 | 42 592F |
| **Central Africa** | **398 320** | **204 238** | | | | |
| Angola | 124 670 | 23 194 | 6 794F | | 2 | 5 830F |
| Botswana | 56 673 | 14 262 | 1 538F | | | 1 443F |
| Burundi | 2 565 | 325 | 4 831 | | | 4 724F |
| Malawi | 9 408 | 3 612 | 10 212F | | | 9 678F |
| Mozambique | 78 409 | 17 357 | 16 041F | | 10 | 15 022F |
| Namibia | 82 329 | 12 569 | | | | |
| Rwanda | 2 467 | 252 | 5 660F | | 4F | 5 392F |
| United Republic of Tanzania | 88 604 | 33 709 | 35 746 | | 7F | 33 626 |
| Zambia | 74 339 | 32 349 | 14 665 | | 29F | 13 465F |
| Zimbabwe | 38 667 | 8 981 | 8 076F | 8 608 | 22 | 6 269F |
| **Tropical South Africa** | **558 131** | **146 609** | | | | |
| Comoros | 223 | 11 | | | | |
| Madagascar | 58 154 | 15 999 | 10 645 | | | 10 179 |
| Mauritius | 203 | 12 | 18 | 9 | | 8 |
| Reunion | 250 | 100 | 36F | 6 | 2 | 31F |
| Saint Helena | 31 | 1 | | | | |
| Seychelles | 27 | 4 | | | | |
| **Insular Africa** | **58 888** | **16 127** | | | | |
| **Tropical Africa** | **2 236 797** | **529 818** | | | | |

| Country | Land Area 1 000 ha | Forest Area 1 000 ha | Roundwood production 1 000 CUM | Roundwood Imports 1 000 CUM | Roundwood Exports 1 000 CUM | Fuelwood+ Charcoal production 1 000 CUM |
|---|---|---|---|---|---|---|
| Algeria | 238 174 | 2 039 | 2 409F | 18 | | 2 081F |
| Egypt | 99 545 | 34 | 2 643F | 414 | 18 | 2 520F |
| Libya Arab Jamahiiya | 175 954 | 400 | 650F | 1 | 1 | 536F |
| Morocco | 71 085 | 3 864 | 1 987F | 430 | | 1 432F |
| Tunisia | 15 536 | 569 | 3 498 | 60 | | 3 291F |
| **North Africa** | **600 294** | **6 905** | | | | |
| Lesotho | 3 035 | 7 | 690F | | | 690F |
| South Africa | 122 104 | 8 208 | 24 218 | 18 | 2 249 | 7 210F |
| Swaziland | 1 720 | 146 | 2 279F | | | 560F |
| **Non tropical South Africa** | **126 859** | **8 361** | | | | |
| **Non tropical Africa** | **727 153** | **15 267** | | | | |
| **Total Africa** | **2 963 950** | **545 085** | **567 729** | **1 033** | **7 285** | **502 252** |
| **Asia / Pacific** | | | | | | |
| Bangladesh | 13 017 | 1 004 | 31 346F | | | 30 620F |
| Bhutan | 4 700 | 2 813 | 1 398F | | | 1 334F |
| India | 297 319 | 64 959 | 293 979F | 285 | 41 | 269 187F |
| Nepal | 13 680 | 5 079 | 20 312F | | | 19 692F |
| Pakistan | 77 088 | 2 023 | 29 873 | 123 | 6F | 27 419 |
| Sri Lanka | 6 463 | 1 885 | 9 493F | 46 | 41 | 8 815F |
| **South Asia** | **412 267** | **77 762** | | | | |
| Combodia | 17 652 | 12 170 | 7 179 | | 329 | 6 512F |
| Laos | 23 080 | 13 177 | 5 094 | | 110 | 4 382F |
| Myanmar | 65 797 | 29 091 | 22 484 | | 982 | 20 040 |
| Thailand | 51 089 | 13 264 | 38 914 | 1 548 | 223 | 36 130F |
| Vietnam | 32 549 | 9 782 | 34 209F | | 215 | 29 810F |
| **Continental S.E.Asia** | **190 167** | **77 484** | | | | |
| Brunei Darussalam | 527 | 458 | 295F | 2 | 2 | 79F |
| Indonesia | 181 157 | 115 674 | 187 089F | 149 | 1 636 | 148 916F |
| Malaysia | 32 855 | 17 664 | 46 037 | 235 | 8 809 | 9 602F |
| Philippines | 29 817 | 8 034 | 38 990 | 464 | 280 | 35 790F |
| Singapore | 61 | 4 | 120F | 322 | 364 | 120F |
| **Insular S.E.Asia** | **244 417** | **141 834** | | | | |
| Papua New Guinea | 45 286 | 36 030 | | | | |
| American Samoa | 20 | 0 | | | | |
| Fiji | 1 827 | 853 | | | | |
| French Polynesia | 366 | 0 | | | | |
| Guam | 55 | 0 | | | | |
| Kiribati | 73 | 0 | | | | |
| New Caledonia | 1 827 | 710 | | | | |
| Niue | 26 | 0 | | | | |
| Pacific Islands | 178 | 0 | | | | |
| Samoa | 283 | 133 | | | | |
| Solomon Islands | 2 799 | 2 410 | | | | |
| Tonga | 72 | 0 | | | | |
| Vanuatu | 1 219 | 809 | | | | |
| **Pacific** | **54 032** | **40 945** | | | | |
| **Total Tropical Asia** | **900 883** | **388 025** | | | | |

| Country | Land Area 1 000 ha | Forest Area 1 000 ha | Roundwood production 1 000 CUM | Roundwood Imports 1 000 CUM | Roundwood Exports 1 000 CUM | Fuelwood+ Charcoal production 1 000 CUM |
|---|---|---|---|---|---|---|
| Afghanistan | 65 209 | 1 199 | 7 251F | | 2 | 5 617F |
| Bahrain | 68 | 0 | | 4 | | |
| Iran | 163 600 | 1 737 | 7 491F | 7 | 5 | 2 549F |
| Iraq | 43 737 | 83 | 155F | 1 | | 105F |
| Jordan | 8 893 | 51 | 12F | 9 | 1 | 8F |
| Kuwait | 1 782 | 5 | | 48 | | |
| Lebanon | 1 023 | 78 | 498F | 67 | | 491F |
| Oman | 21 246 | 0 | | 6 | 8F | |
| Qatar | 1 100 | 0 | | 6 | | |
| Saudi Arabia | 214 969 | 202 | | 67 | 1 | |
| Syrian Arab Republic | 18 420 | 245 | 55F | 17 | 1 | 20F |
| United Arab Emirates | 8 360 | 60 | | 1 319 | 1 | |
| Yemen | 52 797 | 9 | 324F | | | 324F |
| **Middle East** | **601 204** | **3 669** | | | | |
| China | 932 641 | 133 799 | 305 961 | 4 818 | 2 548 | 204 059F |
| Korea, DPR | 12 054 | 6 170 | 4 876F | 50 | 125 | 4 276F |
| Korea, Rep. | 9 902 | 6 291 | 6 485F | 10 807 | 22 | 4 491F |
| Mongolia | 156 650 | 9 406 | 541F | | 43 | 376F |
| **Temperate Asia** | **1 111 247** | **155 666** | | | | |
| **Total Non Tropical Asia** | **1 712 451** | **159 334** | | | | |
| **Total Asia / Pacific** | **2 613 334** | **497 359** | | | | |

## Latin America / Caribean

| Country | Land Area 1 000 ha | Forest Area 1 000 ha | Roundwood production 1 000 CUM | Roundwood Imports 1 000 CUM | Roundwood Exports 1 000 CUM | Fuelwood+ Charcoal production 1 000 CUM |
|---|---|---|---|---|---|---|
| Costa Rica | 5 106 | 1 456 | 4 399F | 16 | | 3 289F |
| El Salvador | 2 085 | 127 | 6 654F | 6 | | 6 508F |
| Guatemala | 10 843 | 4 253 | 16 555 | 4 | | 15 818 |
| Honduras | 11 189 | 4 608 | 6 312F | 1 | 71 | 5 702F |
| Mexico | 190 869 | 48 695 | 22 144 | 42 | 224 | 16 095F |
| Nicaragua | 11 875 | 6 027 | 3 716 | 2 | 1 | 3 531F |
| Panama | 7 599 | 3 123 | 1 053F | 7 | 2 | 935F |
| **Central America** | **239 566** | **68 289** | | | | |
| Barbados | 43 | 0 | 5F | 2 | 5F | |
| Bermuda | 5 | 0 | | 12 | 2F | |
| British Virgin Islands | 15 | 3 | | | | |
| Cayman Islands | 26 | 0 | | 2 | | |
| Monserrat | 10 | 3 | | | | |
| Netherlands Antilles | 80 | 0 | | 5 | | |
| St.Pierre & Miquelon | 23 | 0 | | | | |
| United States Virgin Islands | 34 | 14 | | | | |
| Antigua and Barbuda | 44 | 10 | | | | |
| Bahamas | 1 001 | 186 | 117F | 22 | 2 | |
| Belize | 2 280 | 1 998 | 188F | | 5 | 126F |
| Cuba | 10 982 | 1 960 | 3 146F | | | 2 535F |
| Dominica | 75 | 44 | | | | |
| Dominican Republic | 4 838 | 1 084 | 982F | 13 | | 976F |
| French Guyana | 8 815 | 7 997 | | | | |
| Grenada | 34 | 6 | | 3 | | |
| Guadeloupe | 169 | 93 | 15F | 14 | | 15F |
| Guyana | 19 685 | 18 424 | 180F | | 31 | 14F |
| Haiti | 2 756 | 31 | 6 293F | | | 6 054F |
| Jamaica | 1 083 | 254 | 477 | 19 | 1F | 412 |
| Martinique | 106 | 43 | 12 | 8 | | 10F |
| Puerto Rico | 886 | 324 | | | | |
| St.Kitts and Nevis | 36 | 13 | | 1 | | |

| Country | Land Area 1 000 ha | Forest Area 1 000 ha | Roundwood production 1 000 CUM | Roundwood Imports 1 000 CUM | Roundwood Exports 1 000 CUM | Fuelwood+ Charcoal production 1 000 CUM |
|---|---|---|---|---|---|---|
| St.Lucia | 61 | 5 | | 7 | | |
| St.Vincent | 39 | 11 | | 2 | | |
| Suriname | 15 600 | 14 776 | 112 | | 3 | 19 |
| Trinidad and Tobago | 513 | 168 | 68 | 5 | 1 | 22F |
| **Caribbean** | **69 239** | **47 447** | | | | |
| Bolivia | 108 438 | 49 345 | 2 227 | | 15 | 1 322F |
| Brazil | 845 651 | 566 007 | 275 303F | 5 | 1 674 | 197 400F |
| Colombia | 103 870 | 54 190 | 21 187F | 7 | 22 | 17 504F |
| Ecuador | 27 684 | 12 007 | 7 691 | | 86 | 5 800 |
| Paraquay | 39 730 | 12 868 | 8 592F | 7 | 6 | 5 486F |
| Peru | 128 000 | 68 090 | 12 574 | 1 | 4 | 10 674 |
| Venezuela | 88 205 | 45 943 | 2 276F | 22 | 5 | 1 063F |
| **Tropical South America** | **1 341 578** | **808 450** | | | | |
| **Tropical L.Am / Caribbean** | **1 650 383** | **924 187** | | | | |
| Argentina | 273 669 | 34 436 | 11 683F | 6 | 670 | 5 822F |
| Chile | 74 880 | 8 033 | 33 275 | 3 | 6 531 | 9 972 |
| Uruguay | 17 481 | 813 | 4 087F | 4 | 103 | 3 044F |
| **Non tropical South America** | **366 030** | **43 283** | | | | |
| **Total Latin America / Caribbean** | 2 016 413 | 967 469 | | | | |
| **Total developing countries** | 7 593 697 | 2 009 912 | | | | |
| **Grand Total** | 12 935 867 | 3 442 369 | | | | |

# Global Forest Machine Markets 1956–1996

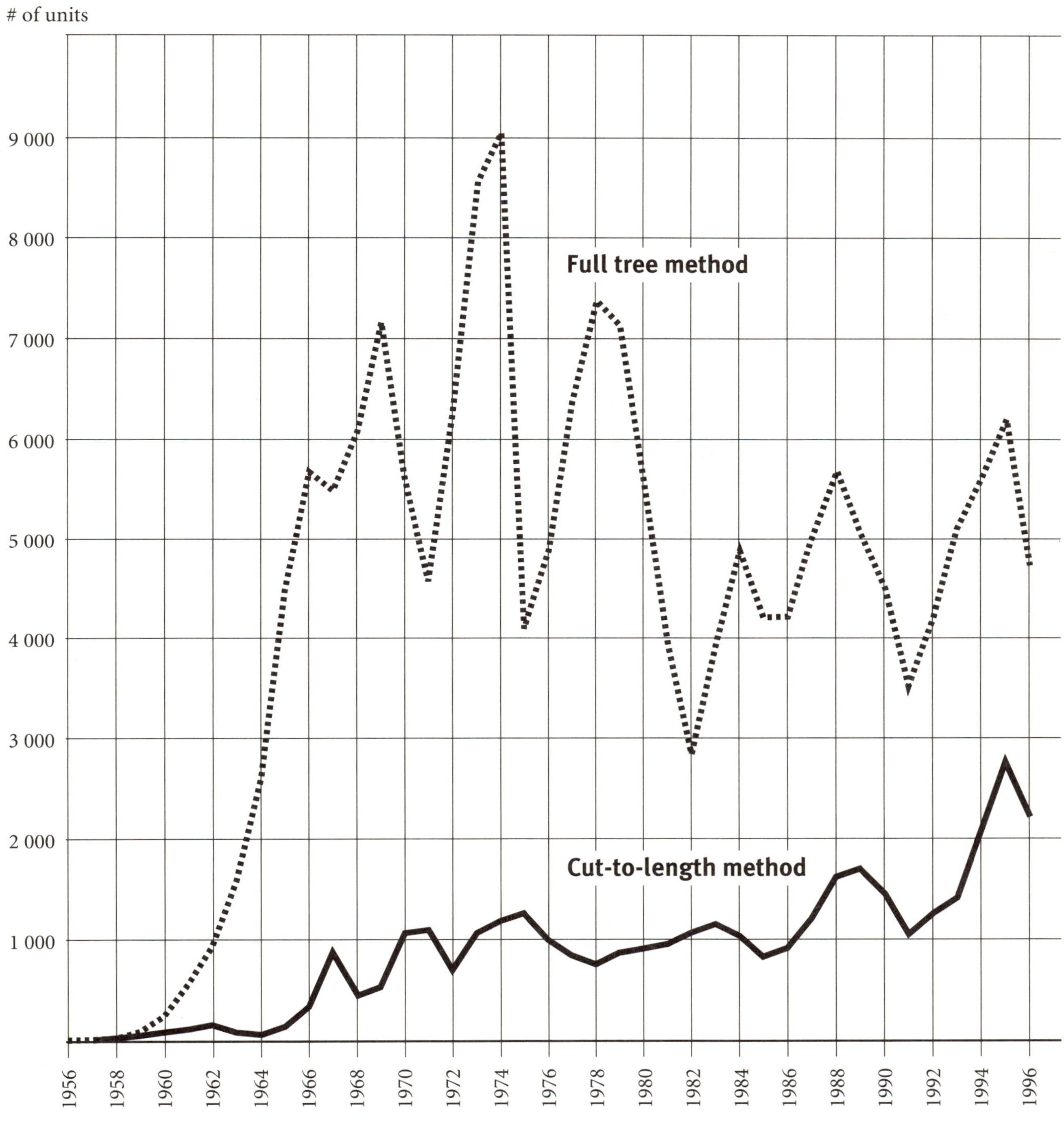


Full tree machines: feller-bunchers, skidders and delimbers
Cut-to-length machines: harvesters, processors and forwarders

Source: Hans-Åke Elofson, Gösta Gustafsson, Mika Laatikainen and Wayne Smith, Timberjack Group.

# Highlights of Forest Machine Evolution

## North America

**1882** John Dolbeer produced the Dolbeer Donkey, a steam-powered log-skidding machine, in California.

**1883** Horace Butters built the first Lidgerwood steam skidder in Michigan.

**1891** Ephraim Shay built his first logging locomotive in Michigan.

**1900** Alvin Lombard built a steam log hauling tractor in Maine driven by the first set of endless tracks.

**1904** The Holt brothers built their first Caterpillar tractor in California.

**1913** The White Motor Company of Ohio produced the first truck designed to haul logs.

**1920** Charles Wolf patented a semi-portable, electric chain saw in Washington.

**1926** Robert LeTourneau invented the dozer blade for use on crawler tractors.

Dolmar and Stihl, German chainsaws, introduced to global markets.

**1949** McCullough introduced its first chain saw, the model 5-49, which captured two-thirds of the North American market within two years.

**1951** First rubber-tired skidder, the Westfall Performer, came on market in Oregon.

## Scandinavia

**1911** Lagerlöf experimented with a mobile electric winch sled.

**1913** Hugo Sandberg tested a Lombard steam hauler.

First Swedish-made farm tractor, Munktell, produced.

**1916** Scandinavia's first chain saws, the Arbor, Gerber and Sektor, appeared.

**1930** Sweden tried out trucks, winches and farm tractors for forest work.

**1936** Sweden's first chain saw logging site opened.

**1942** Sweden started importing Oliver Cletrac half-track tractors from the United States.

**1950** Ösa installed half tracks on a Volvo BM farm tractor.

Erland Andersson developed a debarking machine utilizing compressed air.

## North America

**1953** The Logger's Dream, a cable loader and yarder, developed by Taylor Machine Works in Mississippi.

Big Stick loader introduced by Brown Machinery Service in Mississippi.

**1954** Wagner Tractor produced the first frame-steered skidder, the Loggermobile, in Oregon.

Tom Busch, in Louisiana, developed a hydraulic tree shear to mount on a farm tractor.

The Bonnard Logger, a frame-steered forwarder, produced in Quebec, along with Harrison Pulpwood Harvester and Jarck Go Getter in South Carolina.

Madill built its first portable steel spar in British Columbia.

**1957** Rudy Vit finished prototype feller buncher in Quebec.

**1958** Dwight Garrett, in Washington, built the Tree Farmer, a small frame-steered skidder.

Leo Heikkinen built first North American knuckleboom loader in Wisconsin.

**1959** Busch Combine prototype built in Louisiana by Tom Busch.

**1960** Norval Morey built first mobile debarkers in Michigan.

**1961** Timberland Machines built first Canadian frame-steered skidder, the Timberjack.

**1965** Beloit produces first grapple skidder, the Logging Tractor.

First fully mechanized logging systems, the LogAll feller-skidder and Arbomatik processor, built by Logging Research Associates in Quebec.

**1966** Ray Hahn built his first delimbing and bucking machine, the Hahn Harvester, in Minnesota.

NESCO in Ontario and Nestor Siiro in Minnesota build first single-grip harvesters.

**1968** Erv Drott built a knuckleboom-mounted feller buncher in Wisconsin.

## Scandinavia

**1954** Iggesund launched the Cambio debarking machine.

**1957** Scandinavia's first forest tractor, the Bamse, made a big impact at the Forest Weeks fair.

**1959** Hiab's hydraulic loader appeared.

**1961** Tree Farmer skidder, Vit feller-skidder and Beloit harvester appeared in Sweden.

The VSA, Scandinavia's first forwarder with frame steering completed.

**1964** "The Big Bear" jointly developed by Volvo BM and Ösa started a breakthrough of frame steering.

**1965** Garpenberg Forestry University built Europe's first felling grapple.

**1966** Sweden's first processor, the VSA Brunett, launched.

## North America

**1976** Hydro-Ax marketed hydrostatic drive-to-tree feller buncher.

**1978** Pat Crawford built prototype tracked feller buncher in Wisconsin for selective, steep-slope harvesting.

**1979** Irvin Bell brought three-wheeled feller buncher from South Africa to US and made deal with Morbark to build for North American market.

**1981** Koehring-Waterous built tracked feller buncher in Ontario.

**1982** Tanguay built knuckleboom delimber in Quebec.

Koehring and FERIC developed high speed disk saw in Ontario.

Pull-through delimber developed by CTR and others.

**1985** Peterson introduced a chain-flail debarker in Oregon.

## Scandinavia

**1972** Sakari Pinomäki built the Pika 75 harvester.

**1973** Ösa 710 feller-buncher launched era of harvesters and hydrostatic power transmissions.

First Scandinavian automated measuring systems based on relay technology appeared.

**1978** Rosenberg & Kruuse built the Skogsjan grapple processor invented by Jan Eriksson.

Lars Bruun's Mini-Bruunett pointed the way for small and forest-friendly tractors.

First microprocessor-based measuring systems appeared.

**1983** Silva Nova planting machine built.

The SP 21 began the single-grip harvester era.

**1988** Automated cubic measuring systems introduced.

**1992** Finland approved the automated measuring reports as documentation for timber sales.

**1993** First computer-controlled power transmissions appeared.

**1995** Plustech introduced the Walking Machine.

**1996** Plustech developed a harvesting simulator.

# A Short History of Timberjack's Evolution

## Scandinavia

**1915**
Oy Lokomo Ab is established.

**1918**
Ösa is established in Alfta, Sweden.

**1942**
Rauma-Raahe Oy is established.

**1952**
Rauma-Repola Oy is established.

**1966**
Lokomo signs an agreement to produce forest machines and timber processing facilities with the American company, Beloit.

**1970**
Rauma-Repola acquires Lokomo.

**1971**
Lokomo signs a retail agreement with Hankkija.

**1974**
Lokomo and Ösa start cooperation.

**1974**
Lokomo's forest machine production moves from Tampere to Joensuu.

**1974**
Rauma-Repola acquires 50% of the Vincent machine shop in Quebec.

**1976**
Vincent is closed down.

**1976**
Rauma-Repola talks with Valmet about joining forces.

**1978**
Rauma-Repola makes a decision to continue in forest machine business.

**1979**
Rauma-Repola acquires 49.5% of Ösa.

**1981**
Rauma-Repola acquires 100% of Ösa, and 100% of Forrex from Sponsor which includes Bruun Systems, Makeri and Formacomp.

**1982**
Lokomo dissolves retail agreement with Hankkija.

**1983**
Rauma-Repola establishes Lokomo Forest.

**1984**
Rauma-Repola's forest machine group acquires Kockums' forest machine business.

**1985**
Rauma-Repola sells Makeri and Stensele Mekaniska Värkstad, and closes Kockums Norway and Forrex.

**1986**
Rauma-Repola purchases 60% of the French company, Cemet-Agrip, and signs marketing agreement with Timberjack.

**1988**
Rauma-Repola consolidates forest machine operations under the name FMG, acquires 100% of Cemet-Agrip and signs co-operation agreement with the Japanese company, Komatsu.

## North America

**1844**
Waterous Engine Works Co. is established.

**1947**
Timberland Machines Ltd. is established.

**1953**
Koehring Company purchases Waterous and changes name to Koehring-Waterous Ltd.

**1959**
Ellicott Corporation acquires Timberland Machines and changes name to Timberland-Ellicott Ltd.

**1964**
Hudson Sales is established.

**1965**
Timberland-Ellicott establishes Timberjack Machines Ltd. to manufacture vehicles.

**1967**
Eaton Yale & Towne Corporation acquires Timberjack Machines and changes name to Eaton Yale & Towne Canada Ltd.

**1981**
Company's name is changed to Timberjack.

**1984**
Investment company, Sequoia, and Timberjack's senior management buy Timberjack from Eaton Yale.

**1988**
Timberjack acquires Koehring-Waterous.

**1988**
Timberjack buys Peerless.

**1988**
Timberjack is listed at the New York Stock Exchange.

## The combined companies

**1989**
Rauma-Repola acquires 100% of Timberjack, Koehring and Peerless.

**1990**
Rauma-Repola's forest machine operations reorganized as FMG Timberjack.

**1991**
Rauma-Repola merges with Yhtyneet Paperitehtaat [United Paper Mills]. FMG Timberjack becomes part of the Rauma Group.

**1994**
Name changes to Timberjack Group.

**1996**
Timberjack acquires Hudson Sales.

**1997**
Timberjack celebrates 50th anniversary.

# Sources and Acknowledgments

Material for this book was gathered from a wide variety of sources, primarily in the United States, Canada, Finland and Sweden. Books on the subject which were particularly useful included: Ralph C. Bryant, *Logging* (New York: John Wiley & Sons, 1923); Nelson C. Brown, *Logging* (New York: John Wiley & Sons, 1949); Steve Conway, *Logging Practices* (San Francisco: Miller Freeman, 1976); Ken Drushka, *Working in the Woods* (Madeira Park: Harbour, 1992); Bernhard E. Fernow, *A Brief History of Forestry* (Toronto: University of Toronto, 1911); A. Koroleff, *Pulpwood Cutting* (Montreal: Canadian Pulp & Paper Association, 1941); A. Koroleff, *Logging Mechanization in the U.S.S.R.* (Montreal: Pulp & Paper Research Institute of Canada, 1952); Donald MacKay, *The Lumberjacks* (Toronto: McGraw-Hill Ryerson, 1978); Jaakko Kahiluoto, *Rauma-Repolan metsätaival* (Rauma: Länsi-Suomi, 1990); Henning Hamilton, Pauli Leiwo & Markku Rauhalahti, *Muuttuva metsä* (Helsinki: Suomen Metsäyhdistys, 1988); Kullervo Kuusela, *The Dynamics of Boreal Coniferous Forests* (Helsinki: Sitra, 1990); Juhani Pelkonen, *Nuortin konesavotta 1913–1916* (Tampere: Lapin Metsämuseoyhdistys, 1976); Edmund von Berg, *Kertomus Suomenmaan metsistä 1858* (Jyväskylä: Metsälehti Kustannus, 1995); WWF International and IUCN, *Forests for life* (Surrey: WWF-UK, 1996); Kahala, Kilkki, Niku, Nikunen & Pösö, *Puoli vuosisataa koneellista puunkorjuuta* (Jyväskylä: Gummerus, 1996); K.A.G. Staaf & N.A. Wiksten, *Tree harvesting techniques* (Dordrecht: Martinus Nijhoff Publishers, 1984); Carl Einar Malmberg, *Mekanisering av skogsodling* (Stockholm: STU, 1990); Martin Östberg, *En smedjas förvandling – ÖSAs historia* (Bollnäs: FMG, 1990) Sven Embertsén, *Virkesdrivning inom kramforsdelen av SCA 1911–1965* (Stockholm: Skogshögskolan, 1976); Simo Hannelius & Kullervo Kuusela, *Finland — the Country of Evergreen Forests* (Tampere: XXth IUFRO World Conference, 1995); and Terho A. Könönen, *Lokomo 70 vuotta* (Joensuu: Lokomo, 1985).

Several trade magazines, past and present, provided detailed information on particular logging machines. Preeminent among these publications is *Timber Harvesting*, edited by D.K. Knight in Montgomery, Alabama, whose encouragement and assistance on this project was invaluable. Other helpful trade periodicals include *Truck Logger, Chain Saw Age, The Timberman, Steam Machinery, Skogen, Koneyrittäjäin lehti, Tekniikka ja Talous*, and *Kone, työ ja business.*

Some of the most useful publications were those prepared by organizations devoted to the development of logging equipment, particularly the Woodlands Section of the Canadian Pulp & Paper Association, the American Pulpwood Association, the Swedish Skogforsk and Värmlands Skogsarbetsstudier, and the Finnish section of the WWF. The Food and Agricultural Organization was also most helpful in providing back issues of its papers on forest machinery and current statistical data.

A number of unpublished documents have also contributed greatly to our knowledge. One of these, Tom Walbridge's doctoral dissertation, *The Design of Harvesting Systems and Machines for use in Pulpwood Stands of the Tennessee Valley as Dictated by Intensive Forest Management,* was completed in 1960. Another helpful manuscript written by Ross Silversides, a Canadian forest engineer who died in 1993, is titled *Logging Mechanization in Eastern Canada*, and is planned for publication in the near future. Information was collected from Bjarne Torgny's manuscript and photo archives, *Kort historik övër drivningsteknisk utvecling i Sverige under 50 år*; from Jaakko Piironen and Anders Staaf's personal writings; Kauko Papunen and Gösta Gustafsson's abundant scrapbook and brochure archives; and from Kauko Papunen's detailed report, *Koneellisen puukorjuun kehittyminen Suomessa*, written especially for this book.

Several museums and archives provided generous assistance to this book, including the Forest History Society in Durham, North Carolina; the Campbell River, British Columbia, Museum and Archives; the Canadian Pulp & Paper Association in Montreal; the Special Collections Library at the University of British Columbia; the Southern Forest Heritage Museum & Research Center in Long Leaf, Louisiana; the S.F. Austen State University Library in Nacogdoches, Texas; Skogforsk, Upsala; Skogsmestarsskolan, Skinnskatteberg; Luston Metsämuseo, Punkaharju; Lapin Metsämuseoyhdistys, Rovaniemi; Metsäntutkimuslaitoksen kuva-arkisto, Helsinki; and Metsäkuvapalvelu, Helsinki.

And finally, there is a group of individuals without whom this book could not have been published. In addition to those persons who have been identified in the text, we are enormously grateful for the assistance of: Jim Collins, Bob Lanford, Robin Clark, Juan Carlos Carmona, Bert Reid, Tony Rotherham, Cheryl Oakes, Takao Gen, Alan Auter, Roger Bellshaw, Gar Scott, John Lyons, Bob Cole, Gordon Franklin, John Manz, Ed Mace, Gordon Sims, Rudi Vit, Shozo Sasaki, Myles Mac Donncadha, Neil Ward, D.K. Knight, Walter Jarck, Dave Baird, Leo LeTourneau, Cy Roberts, Nils Hedman, Gustav Widlund, Tore Rautila, Torvald Hallberg, Martin Östberg, Sven Embertzén, Anders Staaf, Carl Einar Malmberg, Allan Mårtenson, Esko T. Helin, Arvo Torvinen, Simo Hannelius, Tauno Tolppa, Jaakko Piironen, Sakari Tiala, Pekka Kokko, Lars Bruun, Matti Kärkkäinen, Dag Myhrman, Bill Sharer, Orvo Tarvainen, Pertti Metso, Harri Karjalainen, Risto Seppälä, Juhani Saloranta, Miikka Marttiini, Raimo Inkiläinen, Matti Sihvonen, Rainer Häggblom, Hannu Hytönen, Tapani Holma and Pasi Järvinen.

We would especially like to thank three veterans in the business: Gösta Gustafsson of Skinskatteberg, Sweden, and Kauko Papunen of Ylöjärvi, Finland, both of whom work for the Timberjack Group, and Tom Walbridge, the former head of the U.S. Harvesting Research Project, all who read the manuscript, selected photographs, offered advice and patiently answered questions. As well, a number of individuals at Timberjack gave us their full support; they include Tor Wilson, Larry Cumming, Tim White, John Dix, Don Small, Donna Bannon, Sylvia Terryberry, Tarja Kivelä, Kati Renvall, Jukka Moisander, Matti Kaarnametsä, Mikko Rysä, Jorma Kiuru, Hans-Åke Elofson, Pertti Huusko, Timo Kyttälä, Mirja Isotalo, Seppo Taatila, Raul Wisotzky, Göran Junevik, Britt Tröen and Karl-Erik Jonsson. We also extend special thanks to Kathleen Kuosmanen of Seattle and Marja Hakola of Helsinki for their translations, Pipsa Löfqvist for collecting information, Tuula Konttinen for analysis and translations from Swedish; and Kati Renvall for managing the project. Dan Francis is to be commended for editing the English version, as is Eija Leskinen for proofing the Finnish text. Finally, we would like to thank the book's designer, Jorma Hinkka, for pulling together many loose pieces into two complete publications.

# The Authors

Ken Drushka is a former journalist who spent fifteen years working in the Canadian forest industry. He has since published several books on the forestry, including *Working in the Woods*, a history of British Columbia coastal logging, and *HR: A biography of H.R. MacMillan.* He writes a monthly column on the forest industry for a business paper in Vancouver.

Hannu Konttinen is a Helsinki advertising man, a freelance writer and illustrator, who has created and designed forest machine advertisements for more than a decade. He has written a number of short publications on light construction technology, soapstone fireplaces and ovens, nature conservation and industrial recruiting.

# Authors' Comments

**"**

*When I was approached by Timberjack to write the English version of a book marking the company's fiftieth anniversary, it somehow struck me as appropriate. I was about to celebrate the thirtieth anniversary of my first job in the forest industry, driving a Timberjack skidder at a haywire gyppo show in the British Columbia coastal rainforest. Just about everything that could possibly go wrong with a logging operation did, including my paycheck. But, in spite of that, I was hooked on the forest industry. Ever since, the business of logging — or anything to do with forests — has seemed to me to be the most interesting subject imaginable.*

*Working on this book has, once again, confirmed my initial reaction. The subject is endlessly fascinating and the people involved are the best anyone could hope to meet. It has also convinced me of something I have long suspected: The strength and future of the forest industry rests ultimately on the creative abilities and fundamental integrity of the people who work in it. That being so, it seems to me the future of this business looks good and I am happy to have a role in it, however small.*

Ken Drushka

**"**

*When I told my acquaintances I was writing a book about forest machines, some of them said: "The forest machine is bad. Write about nature conservation!"*

*I had just written a small guide for Finland's World Wildlife Fund, so I had been redeemed and it felt good being able to defend forest machines.*

*When I was fifteen years old, I worked in the forest for a week in Padasjoki with farmer Helvilä and Harmo the horse. The farmer felled pines with a chain saw, I chopped off the branches with an ax and Harmo skidded the logs to the road. It was hard-working holiday for a schoolboy, but all in a winter day's work for the farmer with his bad back and for Harmo. They took from the forest bread for the family and an education for the children.*

*After this experience, the need for forest machines was clear to me. Since then, I have always wondered how the media managed to get forest machines to symbolize poor forest management. Poor logging practices are the fault of the people operating the machines — not of the machines.*

Hannu Konttinen